PRACTICING TO TAKE THE

GRE® CHEMISTRY TEST

2nd Edition

INCLUDES:

- An actual GRE Chemistry Test administered in 1989-90
- Sample questions, instructions, and answer sheets
- Percent of examinees answering each question correctly

AN OFFICIAL PUBLICATION OF THE GRE BOARD

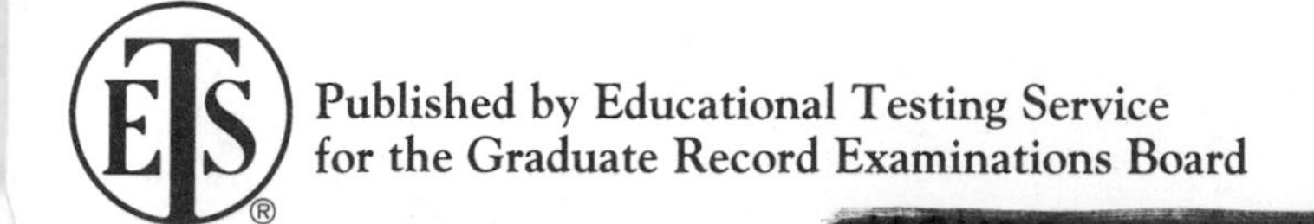

Published by Educational Testing Service
for the Graduate Record Examinations Board

The Graduate Record Examinations Program offers a General Test measuring developed verbal, quantitative, and analytical abilities and Subject Tests measuring achievement in the following 16 fields:

- Biochemistry, Cell and Molecular Biology
- Biology
- Chemistry
- Computer Science
- Economics
- Education
- Engineering
- Geology
- History
- Literature in English
- Mathematics
- Music
- Physics
- Political Science
- Psychology
- Sociology

The tests are administered by Educational Testing Service under policies determined by the Graduate Record Examinations Board, an independent board affiliated with the Association of Graduate Schools and the Council of Graduate Schools.

The Graduate Record Examinations Board has officially made available for purchase practice books, each containing a full-length test, for 15 of the Subject Tests. A practice book is not available for the Biochemistry, Cell and Molecular Biology Test at this time. Two General Test practice books are also available. These practice books may be purchased by using the order form on page 119.

Individual booklets describing each test and including sample questions are available free of charge for all 16 Subject Tests. These booklets may be requested by writing to:

Graduate Record Examinations
Educational Testing Service
P.O. Box 6014
Princeton, NJ 08541-6014

In association with Warner Books, Inc., a Warner Communications Company.

USA: 0-446-39201-4
CAN: 0-446-39202-2

TABLE OF CONTENTS

BACKGROUND FOR THE TEST

TAKING THE TEST

BACKGROUND FOR THE TEST

PRACTICING TO TAKE THE GRE® CHEMISTRY TEST

This practice book has been published on behalf of the Graduate Record Examinations Board to help potential graduate students prepare to take the GRE Chemistry Test. The book contains the actual GRE Chemistry Test administered in October 1989, along with a section of sample questions, and includes information about the purpose of the GRE Subject Tests, a detailed description of the content specifications for the GRE Chemistry Test, and a description of the procedures for developing the test. All test questions that were scored have been included in the practice test.

The sample questions included in this practice book are organized by content category and represent the types of questions included in the test. The purpose of these questions is to provide some indication of the range of topics covered in the test as well as to provide some additional questions for practice purposes. These questions do not represent either the length of the actual test or the proportion of actual test questions within each of the content categories.

Before you take the full-length test, you may want to answer the sample questions. A suggested time limit is provided to give you a rough idea of how much time you would have to complete the sample questions if you were answering them on an actual timed test. After answering the sample questions, evaluate your performance within content categories to determine whether you would benefit by reviewing certain courses.

This practice book contains a complete test book, including the general instructions printed on the back cover and inside back cover. When you take the test at the test center, you will be given time to read these instructions. They show you how to mark your answer sheet properly and give you advice about guessing.

Try to take this practice test under conditions that simulate those in an actual test administration. Use the answer sheets provided on pages 111 to 118 and mark your answers with a No. 2 (soft-lead) pencil as you will do at the test center. Give yourself 2 hours and 50 minutes in a quiet place and work through the test without interruption, focusing your attention on the questions with the same concentration you would use in taking the test to earn a score. Since you will not be permitted to use them at the test center, do not use keyboards, dictionaries or other books, compasses, pamphlets, protractors, highlighter pens, rulers, slide rules, calculators (including watch calculators), stereos or radios with headphones, watch alarms including those with flashing lights or alarm sounds, or paper of any kind.

After you complete the practice test, use the work sheet and conversion tables on pages 66 and 67 to score your test. The work sheet also shows the estimated percent of GRE Chemistry Test examinees from a recent three-year period who answered each question correctly. This will enable you to compare your performance on the questions with theirs. Evaluating your performance on the actual test questions as well as the sample questions should help you determine whether you would benefit further by reviewing certain courses before taking the test at the test center.

We believe that if you use this practice book as we have suggested, you will be able to approach the testing experience with increased confidence.

ADDITIONAL INFORMATION

If you have any questions about any of the information in this book, please write to:

Graduate Record Examinations
Educational Testing Service
P.O. Box 6000
Princeton, NJ 08541-6000

PURPOSE OF THE GRE SUBJECT TESTS

The GRE Subject Tests are designed to help graduate school admission committees and fellowship sponsors assess the qualifications of applicants in their subject fields. The tests also provide students with an assessment of their own qualifications.

Scores on the tests are intended to indicate students' knowledge of the subject matter emphasized in many undergraduate programs as preparation for graduate study. Since past achievement is usually a good indicator of future performance, the scores are helpful in predicting students' success in graduate study. Because the tests are standardized, the test scores permit comparison of students from different institutions with different undergraduate programs.

The Graduate Record Examinations Board recommends that scores on the Subject Tests be considered in conjunction with other relevant information about applicants. Because numerous factors influence success in graduate school, reliance on a single measure to predict success is not advisable. Other indicators of competence typically include undergraduate transcripts showing courses taken and grades earned, letters of recommendation, and GRE General Test scores.

DEVELOPMENT OF THE GRE CHEMISTRY TEST

Each new edition of the Chemistry Test is developed by a committee of examiners composed of professors in the subject who are on undergraduate and graduate faculties in different types of institutions and in different regions of the United States. In selecting members for the committee of examiners, the GRE Program seeks the advice of the American Chemical Society.

The content and scope of each test are specified and reviewed periodically by the committee of examiners who, along with other faculty members who are also subject-matter specialists, write the test questions. All questions proposed for the test are reviewed by the committee and revised as necessary. The accepted questions are assembled into a test in accordance with the content specifications developed by the committee of examiners to ensure adequate coverage of the various aspects of the field and at the same time to prevent overemphasis on any single topic. The entire test is then reviewed and approved by the committee.

Subject-matter and measurement specialists on the ETS staff assist the committee of examiners, providing information and advice about methods of test construction and helping to prepare the questions and assemble the test. In addition, they review every test question to identify and eliminate language, symbols, or content considered to be potentially offensive, inappropriate, or serving to perpetuate any negative attitudes. The test as a whole is also reviewed to make sure that the test questions, where applicable, include an appropriate balance of people in different groups and different roles.

Because of the diversity of undergraduate curricula in chemistry, it is not possible for a single test to cover all the material an examinee may have studied.

The examiners, therefore, select questions that test the basic knowledge and understanding most important for successful graduate study in the field. The committee keeps the test up-to-date by regularly developing new editions and revising existing editions. In this way, the test content changes steadily but gradually, much like most curricula.

When a new edition is introduced into the program, it is equated; that is, the scores are related by statistical methods to scores on previous editions so that scores from all editions in use are directly comparable. Although they do not contain the same questions, all editions of the Chemistry Test are constructed according to equivalent specifications for content and level of difficulty, and all measure equivalent knowledge and skills.

After a new edition of the Chemistry Test is first administered, examinees' responses to each test question are analyzed to determine whether the question functioned as expected. This analysis may reveal that a question is ambiguous, requires knowledge beyond the scope of the test, or is inappropriate for the group or a particular subgroup of examinees taking the test. Such questions are not counted in computing examinees' scores.

CONTENT OF THE GRE CHEMISTRY TEST

The test consists of about 150 multiple-choice questions, some of which are grouped in sets and based on such materials as a descriptive paragraph or experimental results. From the five optional answers provided for each question, students are to choose the one that is correct or best. A periodic table is printed in the test booklets. Test questions are constructed to simplify mathematical manipulations. As a result, neither calculators nor tables of logarithms are needed. If the solution of a problem requires the use of logarithms, the necessary values are included with the question.

The content of the test emphasizes the four fields into which chemistry has been traditionally divided and some interrelationships among the fields. Because of these interrelationships, it is impossible to categorize precisely all questions as testing one and only one field of chemistry. For some candidates a particular question may seem to be associated with one field, whereas other candidates may have encountered the same material in a different field. For example, the knowledge necessary to answer some questions classified as testing organic chemistry may well have been acquired in analytical chemistry courses by some candidates. Consequently, the relative emphasis on the four fields indicated in the following outline of material covered by the test should not be considered definitive.

I. ANALYTICAL CHEMISTRY 15 percent

A. Classical Quantitative Analysis

Titrimetry; separations, including theory and applications of chromatography as well as gravimetry; data handling, including statistical tests (t, F, Q, chi-square); standards and standardization techniques

B. Instrumental Analysis

Basic electronics; electrochemical methods; spectroscopic methods; chromatographic methods

II. INORGANIC CHEMISTRY 25 percent

A. Basic Descriptive Chemistry of the Elements

General chemistry, properties, preparations, reactions of representative elements, transition metals, lanthanides and actinides; behavior as related to periodic trends and electronic structures

B. Periodic and Family Trends

Size, ionization energy, electron affinity, period 2 elements *vs.* period 3 elements, oxidation states, electronic configurations, reactivity, melting and boiling points, magnetic properties

C. Electronic and Nuclear Structure

Atomic theory, fundamental particles, atomic structure, electronic configuration; isotopes, abundances and stabilities of nuclei; relationship to reactivity, spectra, and periodic classifications

D. Transition Metal/Coordination Chemistry

Synthesis, reactions, structure, theory, reaction mechanisms, spectra, bonding, isomerism; organometallic compounds (synthesis, structure, bonding, reaction pathways, catalytic processes)

E. Special Topics

Theories of acid and basis, inorganic polymers, ion separations, ionic solids

III. ORGANIC CHEMISTRY 30 percent

A. Conversions of Functional Groups

Alkenes, acetylenes, dienes, alkyl halides, alcohols, aldehydes and ketones, carboxylic acids and derivatives, amines, aromatic compounds

B. Reactive Intermediates and Reaction Mechanisms

Nucleophilic displacement reactions, eliminations, carbonium ions and other cations, benzyne, reactions involving carbene or dichlorocarbene, carbonyl group reactions, radical reactions, tools of mechanism chemists

C. Molecular Structure

Isomerism and related topics, bonding, stereochemistry, determination by chemical and spectral methods

D. Special Topics

Special reagents, rearrangements, structural features of natural products, synthesis problems, separations, and chromatography

IV. PHYSICAL CHEMISTRY 30 percent

A. General Chemistry

Acid-base theories, colligative properties, gas laws, electrochemistry

B. Classical and Statistical Thermodynamics

Variables: moles, pressure, volume, temperature, heat and work, heat capacity, energy, entropy, Gibbs energy; equations of state; first, second, and third laws; phase, chemical and electrochemical equilibria; Boltzmann's relation, partition functions

C. Quantum and Structural Chemistry

Nuclei, atomic structure, molecular and crystal structure, quantization of energy, basic principles of spectroscopy

D. Kinetics

Kinetic theory of gases, chemical kinetics, collision theory, absolute rate theory, rate laws, reaction mechanisms

Each form of the test samples widely among the topics listed above, but questions on all the topics cannot be included in every edition of the test. Because undergraduate chemistry curricula vary considerably from institution to institution, few candidates will have encountered all of the topics listed. Consequently, no candidate should expect to be able to answer all questions on the edition of the test he or she takes. In fact, an excellent scaled score can be achieved with successful performance on less than 70 percent of the questions.

Successful performances by graduate students of chemistry require a number of talents and skills as well as knowledge. However, studies at a number of departments of chemistry have shown that the best single predictor of success in graduate chemical education is the GRE Chemistry Test score. Understanding of principles and knowledge of facts are both important for obtaining a good score. Candidates in previous years have found it useful to review the texts for their major chemistry courses, including their freshman text, to refresh their memories of basic information about the four fields.

When taking the test, examinees are urged to read a question completely, consider what is required, and identify possible mathematical simplifications that would assist in answering it before deciding whether or not to omit the question.

SAMPLE QUESTIONS

The sample questions included in this practice book are organized by content category and represent the types of questions included in the test. The purpose of these questions is to provide some indication of the range of topics covered in the test as well as to provide some additional questions for practice purposes. **These questions do not represent either the length of the actual test or the proportion of actual test questions within each of the content categories.** A time limit of 170 minutes is suggested to give you a rough idea of how much time you would have to complete the sample questions if you were answering them on an actual timed test. Correct answers to the sample questions are listed on page 63.

When you take the actual GRE test, you will be instructed to mark your answers on the separate answer sheet. The directions for the sample questions have been modified. For these questions, you may record your answers in one of two ways: 1) you can use the option bubbles at the bottom-right corner of each question or 2) you can use one of the sample answer sheets provided in this book.

Material in the following table may be useful in answering the sample questions.

PERIODIC CHART OF THE ELEMENTS

1 H 1.0079																	2 He 4.003
3 Li 6.941	4 Be 9.012											5 B 10.81	6 C 12.011	7 N 14.007	8 O 16.00	9 F 19.00	10 Ne 20.179
11 Na 22.99	12 Mg 24.30											13 Al 26.98	14 Si 28.09	15 P 30.974	16 S 32.06	17 Cl 35.453	18 Ar 39.948
19 K 39.10	20 Ca 40.08	21 Sc 44.96	22 Ti 47.90	23 V 50.94	24 Cr 52.00	25 Mn 54.94	26 Fe 55.85	27 Co 58.93	28 Ni 58.70	29 Cu 63.55	30 Zn 65.38	31 Ga 69.72	32 Ge 72.59	33 As 74.92	34 Se 78.96	35 Br 79.90	36 Kr 83.80
37 Rb 85.47	38 Sr 87.62	39 Y 88.91	40 Zr 91.22	41 Nb 92.91	42 Mo 95.94	43 Tc (97)	44 Ru 101.1	45 Rh 102.91	46 Pd 106.4	47 Ag 107.868	48 Cd 112.41	49 In 114.82	50 Sn 118.7	51 Sb 121.75	52 Te 127.60	53 I 126.90	54 Xe 131.30
55 Cs 132.91	56 Ba 137.33	57 *La 138.91	72 Hf 178.49	73 Ta 180.95	74 W 183.85	75 Re 186.21	76 Os 190.2	77 Ir 192.2	78 Pt 195.09	79 Au 196.97	80 Hg 200.59	81 Tl 204.37	82 Pb 207.2	83 Bi 208.98	84 Po (209)	85 At (210)	86 Rn (222)
87 Fr (223)	88 Ra (226)	89 †Ac (227)															

*Lanthanum Series

58 Ce 140.12	59 Pr 140.91	60 Nd 144.24	61 Pm (145)	62 Sm 150.4	63 Eu 152.0	64 Gd 157.25	65 Tb 158.93	66 Dy 162.50	67 Ho 164.93	68 Er 167.26	69 Tm 168.93	70 Yb 173.04	71 Lu 174.97

†Actinium Series

90 Th 232.0	91 Pa 231.0	92 U 238.03	93 Np 237.0	94 Pu (244)	95 Am (243)	96 Cm (247)	97 Bk (247)	98 Cf (251)	99 Es (252)	100 Fm (257)	101 Md (258)	102 No (259)	103 Lr (260)

I. Analytical Chemistry

Directions: Each of the questions or incomplete statements below is followed by five suggested answers or completions. Select the one that is best in each case.

1. $$2\,H_2O \rightleftarrows H_3O^+ + OH^-$$

   The equilibrium constant for the autoprotolysis of water as given above is 1.00×10^{-14} at 25° C. The pH of a 1.00×10^{-10}-molar HCl solution at 25° C is closest to which of the following?

   (A) 12 (B) 10.5 (C) 10 (D) 7 (E) 2

   Ⓐ Ⓑ Ⓒ Ⓓ Ⓔ

2. In the titration of a monoprotic acid with a monobasic base, the equivalence point is necessarily the point at which

   (A) the midpoint of the color-change interval of the indicator is reached
   (B) the mixture is electrically neutral
   (C) equal numbers of equivalents of the acid and the base have been mixed
   (D) the value of the pH of the solution equals the value of the pK for the acid
   (E) the pH of the mixture is 7

   Ⓐ Ⓑ Ⓒ Ⓓ Ⓔ

3. For a species that is distributed between an aqueous sample and an organic extraction phase, the quantity extracted by the organic phase is

   (A) larger if the total volume of the organic phase is used in portions rather than all at once
   (B) dependent only on the total volume of the organic phase used
   (C) less if ten extractions are made than if five extractions are made
   (D) maximal if the species is weakly acidic and an alkaline aqueous phase is used
   (E) less efficient at a higher temperature if the solubility of the species in both layers remains unchanged

   Ⓐ Ⓑ Ⓒ Ⓓ Ⓔ

4. A sample of uranium-bearing material weighing 1.6000 grams yielded 0.4000 gram of U_3O_8 (molecular weight 842.2). The percentage of uranium (atomic weight 238.1) in the sample is

(A) $\frac{0.4000}{1.6000} \times 100$

(B) $\frac{0.4000}{1.6000} \times \frac{238.1}{842.2} \times 100$

(C) $\frac{0.4000}{1.6000} \times \frac{842.2}{3 \times 238.1} \times 100$

(D) $\frac{0.4000}{1.6000} \times \frac{3 \times 238.1}{842.2} \times 100$

(E) none of the above

Ⓐ Ⓑ Ⓒ Ⓓ Ⓔ

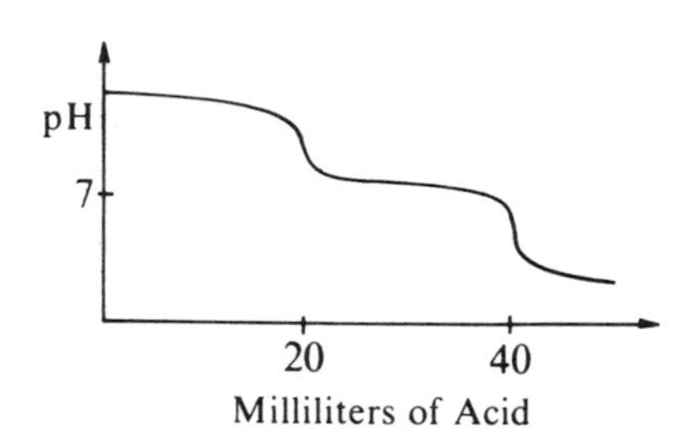

5. When a sample that may have contained NaOH, $NaHCO_3$, and/or Na_2CO_3 was titrated, the titration curve above was obtained. The sample probably contained

(A) pure NaOH
(B) pure Na_2CO_3
(C) pure $NaHCO_3$
(D) NaOH and Na_2CO_3
(E) Na_2CO_3 and $NaHCO_3$

Ⓐ Ⓑ Ⓒ Ⓓ Ⓔ

6. A certain monobasic acid is found to be soluble in aqueous sodium carbonate but insoluble in aqueous sodium bicarbonate. If the ionization constants for carbonic acid are 3×10^{-7} and 6×10^{-11}, the ionization constant for the acid must be

(A) approximately 6×10^{-11}

(B) approximately 3×10^{-7}

(C) greater than 3×10^{-7}

(D) smaller than 6×10^{-11}

(E) greater than 6×10^{-11} but smaller than 3×10^{-7}

Ⓐ Ⓑ Ⓒ Ⓓ Ⓔ

7. Which of the following instruments can be used to carry out a complete analytical evaluation of a metallic alloy?

(A) Polarimeter
(B) Refractometer
(C) Emission spectrograph
(D) Infrared spectrophotometer
(E) Electron microscope

Ⓐ Ⓑ Ⓒ Ⓓ Ⓔ

8. A given solution contains Fe^{3+}, Cr^{3+}, Zn^{2+}, K^{+}, and Cl^{-}. Addition of excess NaOH solution results in

(A) no change in the appearance of the solution
(B) no precipitation and no effervescence
(C) the precipitation of a red-brown substance and the formation of a green solution
(D) the precipitation of a green substance and the formation of a brown solution
(E) the precipitation of a white substance and the formation of a green solution

Ⓐ Ⓑ Ⓒ Ⓓ Ⓔ

9. For aqueous solutions, Raman measurements are usually preferable to the corresponding infrared measurements (2-15 microns) because

(A) scattering problems due to tiny amounts of suspended impurities in the sample are much more severe in the infrared
(B) Raman measurements are more sensitive and permit much smaller concentrations to be determined
(C) Raman measurements are easily capable of much greater resolution than infrared because shorter wavelengths are used
(D) few water-soluble substances absorb in the infrared
(E) water has absorption bands in the 2-15 micron region and comparatively small portions of the spectrum are left free for measurements of solutes

10. The standard potential for the reaction $Q^{3+} + e^- = Q^{2+}$ is E^0_Q. That for $R^+ + e^- = R$ is E^0_R. A solution 1 molar in Q^{3+}, 1 molar in Q^{2+}, and 1 molar in R^+ is agitated in the presence of excess solid R metal. When equilibrium is attained, the solution is found by direct analysis to be 1.3 molar in R^+, 1.3 molar in Q^{2+}, and 0.7 molar in Q^{3+}. On the basis of these data, which of the following conclusions about the standard potentials can be deduced?

(A) $E^0_Q < E^0_R$
(B) $E^0_Q = E^0_R$
(C) $E^0_Q > E^0_R$
(D) $E^0_R > 0$
(E) $E^0_R + E^0_Q = 0$

Ⓐ Ⓑ Ⓒ Ⓓ Ⓔ

11. A molecule is known to absorb radiation at a frequency of 640 wave numbers. The instrument one would probably use to study this absorption band would be

(A) an infrared spectrophotometer
(B) a vacuum ultraviolet spectrometer
(C) a visible spectrometer
(D) a gamma ray spectrometer
(E) a nuclear magnetic resonance spectrometer

12. Liquid column chromatography on alumina would separate *p*-nitrotoluene and *p*-nitroaniline

(A) only slightly since the molecules are nearly the same size
(B) with *p*-nitroaniline eluted first since the stationary phase has little affinity for polar compounds
(C) with *p*-nitrotoluene eluted first since the stationary phase adsorbs polar compounds strongly
(D) with *p*-nitrotoluene eluted first since it has a lower boiling point
(E) with *p*-nitroaniline eluted first since equilibria in the moving phase involve ions

13. The percentage of phosphorus (atomic weight 31.0) in a sample of phosphor bronze may be determined gravimetrically according to the following procedure: The sample is dissolved in *aqua regia*; the solution is diluted slightly with water and is digested at 90° C; the phosphate is precipitated as $MgNH_4PO_4 \cdot 6\ H_2O$ (molecular weight 245.3), and the precipitate is filtered, ignited at 1,000° C to $Mg_2P_2O_7$ (molecular weight 222.6), and weighed. The percentage of phosphorus in the sample is calculated from which of the following?

(A) $\dfrac{(\text{Wt. } Mg_2P_2O_7/245.3)(31.0)}{(\text{Wt. sample})} \times 100$

(B) $\dfrac{(\text{Wt. } Mg_2P_2O_7/222.6)2(31.0)}{(\text{Wt. sample})} \times 100$

(C) $\dfrac{(\text{Wt. } Mg_2P_2O_7/222.6)(31.0/245.3)}{(\text{Wt. sample})} \times 100$

(D) $\dfrac{(\text{Wt. } Mg_2P_2O_7/222.6)(31.0)}{2(\text{Wt. sample})} \times 100$

(E) $\dfrac{(\text{Wt. } Mg_2P_2O_7/245.3)2(31.0)}{(\text{Wt. sample})} \times 100$

14. Which of the following is a likely method for detecting a picogram (10^{-12} gram) of bismuth in an alloy?

(A) Spectrophotometry
(B) Neutron activation analysis
(C) Polarography
(D) Potentiometric titration
(E) Fluorimetry

Ⓐ Ⓑ Ⓒ Ⓓ Ⓔ

15. In a solution containing both triaminotriethylamine (trien) and ethylenediaminetetraacetic acid (EDTA), at pH 10.0 the principal Cu^{2+} species is a trien complex, but at pH 6.0 it is an EDTA complex. The most complete explanation is that

(A) there are fewer side reactions at the lower pH
(B) the anionic EDTA has a greater activity-coefficient correction at higher pH values
(C) the effective stability constant of the Cu-EDTA complex decreases with increasing pH
(D) the ratio of the effective stability constant of the Cu-EDTA complex to that of the Cu-trien complex increases with decreasing pH
(E) EDTA is a hexadentate chelating agent whereas trien is only tetradentate

Ⓐ Ⓑ Ⓒ Ⓓ Ⓔ

16. Desirable properties of a solvent for the nonaqueous titration of a weak acid include all of the following EXCEPT

(A) a low autoprotolysis constant
(B) good proton-accepting ability
(C) adequate solubility of the acid
(D) a moderate dielectric constant
(E) low vapor pressure

Ⓐ Ⓑ Ⓒ Ⓓ Ⓔ

17. If the solution is not supersaturated, what is the maximum Bi^{3+} concentration that can be reached without causing precipitation in a solution containing 0.01 mole per liter of S^{2-} ? (The K_{sp} of bismuth sulfide may be taken as 1×10^{-70}.)

(A) 1×10^{-72} M
(B) 1×10^{-68} M
(C) 9×10^{-64} M
(D) 1×10^{-34} M
(E) 1×10^{-32} M

Ⓐ Ⓑ Ⓒ Ⓓ Ⓔ

18. Which of the following statements about the behavior of glycine during electrophoresis is correct? (For glycine, pK_1 is 2.35 and pK_2 is 9.47.)

(A) In a buffer of pH 1.00, glycine migrates toward the positive electrode.
(B) In a buffer of pH 1.00, glycine does not migrate.
(C) In a buffer of pH 5.91, glycine does not migrate.
(D) In a buffer of pH 5.91, glycine migrates toward the negative electrode.
(E) In a buffer of pH 11.00, glycine migrates toward the negative electrode.

Ⓐ Ⓑ Ⓒ Ⓓ Ⓔ

19.

$$[HA]_{CHCl_3}/[HA]_{H_2O}$$

The value for the distribution constant for an organic acid HA between chloroform and water is 10^2. The dissociation constant of the acid in water is 10^{-5}, and it can be assumed that it does not dissociate or dimerize in the chloroform phase. Increasing the pH of the aqueous phase from 4 to 9 has what effect on the fraction of HA extracted into the chloroform phase?

(A) The fraction is more than doubled.
(B) The fraction is increased but is not doubled.
(C) The fraction is unaffected.
(D) The fraction is decreased but is not halved.
(E) The fraction is reduced to less than half.

Ⓐ Ⓑ Ⓒ Ⓓ Ⓔ

20. Methods of detection applied in gas chromatography include which of the following?

    I. Thermal conductivity
    II. Electron capture
    III. Flame ionization

    (A) I only
    (B) I and II only
    (C) I and III only
    (D) II and III only
    (E) I, II, and III

Ⓐ Ⓑ Ⓒ Ⓓ Ⓔ

21. It is desired to remove zinc ions by means of ion exchange from a sodium chloride solution that has been acidified to a pH of 2.0. Which of the following ion exchange resins is best suited for this application?

    (A) $\left(-R-\overset{H}{\underset{|}{C}}-R- \right)_n$ with para-SO_3H phenyl group
    (B) $\left(-R-\overset{H}{\underset{|}{C}}-R- \right)_n$ with para-$COOH$ phenyl group
    (C) $\left(-R-\overset{H}{\underset{|}{C}}-R- \right)_n$ with para-$NH_3^+Cl^-$ phenyl group
    (D) $\left(-R-\overset{H}{\underset{|}{C}}-R- \right)_n$ with para-$N(CH_3)_3^+Cl^-$ phenyl group
    (E) $\left(-R-\overset{H}{\underset{|}{C}}-R- \right)_n$ with para-OH phenyl group

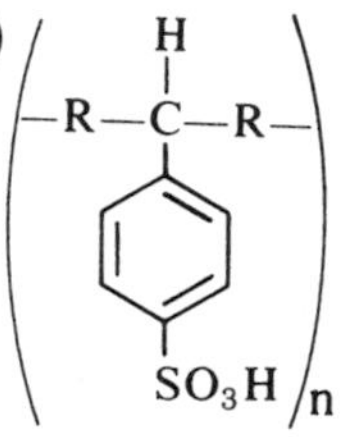

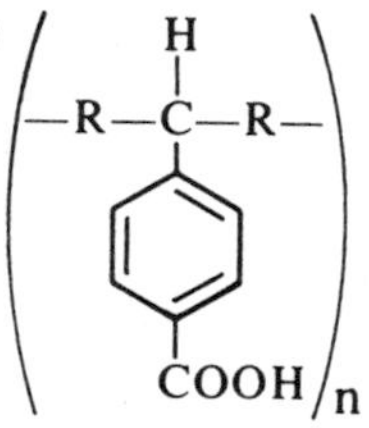

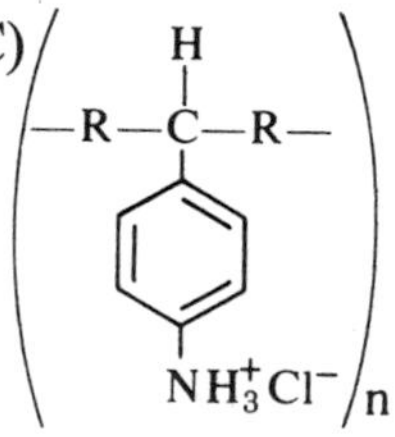

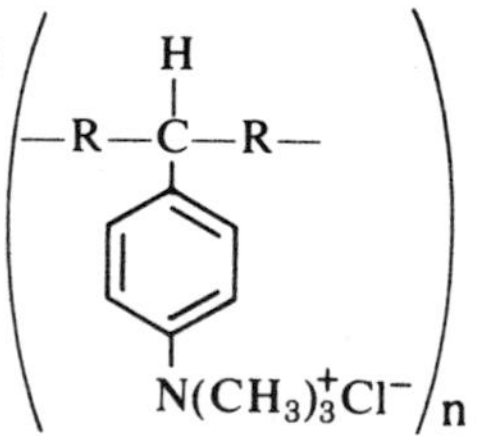

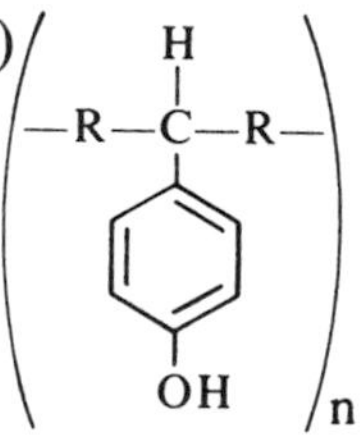

22. Which of the following samples of reducing agents is chemically equivalent to 25 milliliters of 0.20-normal $KMnO_4$ to be reduced to Mn^{2+} and H_2O ?

(A) 100. ml of 0.10 M $H_2C_2O_4 \cdot 2\ H_2O$ to be oxidized to CO_2 and H_2O
(B) 50. ml of 0.10 M H_3AsO_3 to be oxidized to H_3AsO_4
(C) 25 ml of 0.20 M H_2O_2 to be oxidized to H^+ and O_2
(D) 25 ml of 0.10 M $SnCl_2$ to be oxidized to Sn^{4+}
(E) 25 ml of 0.10 M $FeSO_4 \cdot 7\ H_2O$ to be oxidized to Fe^{3+}

23. $Ag_2SO_4(s) \rightleftarrows 2\ Ag^+ + SO_4^{2-}$

A sample of solid silver sulfate was put into a solution originally 0.020 molar in Na_2SO_4. After equilibrium was attained, the concentration of the silver ion in the supernatant liquid was 0.040 molar. If the only significant equilibrium reaction is the one above, what is the value of the solubility product constant of silver sulfate?

(A) 1.6×10^{-3}
(B) 8.0×10^{-4}
(C) 2.6×10^{-4}
(D) 1.3×10^{-4}
(E) None of the above

II. Inorganic Chemistry

Directions: Each of the questions or incomplete statements below is followed by five suggested answers or completions. Select the one that is best in each case.

24. Covalent bonds are most likely to exist between

(A) atoms of nonmetals
(B) atoms of metals
(C) an atom of a nonmetal and an atom of a metal
(D) atoms of elements with great differences in electronegativity
(E) atoms of elements with great differences in ionization energy

Ⓐ Ⓑ Ⓒ Ⓓ Ⓔ

25. Which of the following statements is NOT true of transition elements?

(A) They are all metals.
(B) Most of their ions are colored.
(C) Few of them form complexes.
(D) Most of them contain unpaired electrons.
(E) Most of them have incompletely filled orbitals.

Ⓐ Ⓑ Ⓒ Ⓓ Ⓔ

26. Which of the following halides CANNOT function as a Lewis acid?

(A) $SnCl_4$ (B) $SbCl_5$ (C) BF_3 (D) CCl_4 (E) SiF_4

Ⓐ Ⓑ Ⓒ Ⓓ Ⓔ

27. Which of the following substances has the greatest ionic character in its bonds?

(A) LiCl (B) RbCl (C) $BeCl_2$ (D) $CaCl_2$ (E) BCl_3

Ⓐ Ⓑ Ⓒ Ⓓ Ⓔ

28.

	$E^\circ_{reduction}$		First Ionization Energy
Li^+	–3.045 V	Li	5.4 eV
Na^+	–2.714 V	Na	5.1 eV
Rb^+	–2.925 V	Rb	4.2 eV

Lithium has a higher first ionization energy than rubidium has. Hence, lithium atoms would be expected to form ions less readily than rubidium atoms do. In water, however, lithium atoms are more easily oxidized than rubidium atoms because

(A) Rb is the better reducing agent
(B) Rb is more easily oxidized
(C) Li^+ is more easily reduced
(D) the small Li^+ ions bond strongly to water
(E) Li has a higher melting point than Rb has

Ⓐ Ⓑ Ⓒ Ⓓ Ⓔ

29. The element fluorine cannot be prepared by the oxidation of the fluoride ion with a chemical oxidizing agent in aqueous solution because

(A) the fluoride ion is too strongly hydrated
(B) the fluoride ion is too unstable
(C) fluorine is a very strong oxidizing agent
(D) fluorine is the strongest reducing agent known
(E) the fluorine atom has a very high ionization energy

Ⓐ Ⓑ Ⓒ Ⓓ Ⓔ

30. Which of the following is NOT a direct product of the electrolysis of aqueous sodium chloride or the reaction of one of these products with water?

(A) NaH (B) NaOH (C) HOCl (D) H_2 (E) Cl_2

Ⓐ Ⓑ Ⓒ Ⓓ Ⓔ

31. A 0.1-molar solution of which of the following substances is most acidic?

(A) KNO_3 (B) KCN (C) $AlCl_3$ (D) $NaC_2H_3O_2$
(E) $Cu(C_2H_3O_2)_2$

Ⓐ Ⓑ Ⓒ Ⓓ Ⓔ

32. Which of the following commercial chemicals is most widely used as a paint pigment?

(A) Al_2O_3 (B) MgO (C) SiO_2 (D) CaO (E) TiO_2

Ⓐ Ⓑ Ⓒ Ⓓ Ⓔ

33. One should expect which of the following to have the highest melting point?

(A) NaCl (B) BCl_3 (C) $AlCl_3$ (D) S_2Cl_2 (E) $SeCl_4$

Ⓐ Ⓑ Ⓒ Ⓓ Ⓔ

34. Which of the following statements best describes the relative acidities in liquid ammonia of hydrochloric acid and acetic acid?

(A) Both appear to be weak acids.
(B) Both appear to be strong acids.
(C) Acetic acid appears to be a stronger acid than HCl is.
(D) HCl appears to be a weak acid and acetic acid appears not to be an acid.
(E) Neither shows acidic behavior.

Ⓐ Ⓑ Ⓒ Ⓓ Ⓔ

35. In a Born-Haber cycle which of the following is most important in accounting for the highly negative standard heat of formation for MgO, a stable ionic compound?

(A) Dissociation energy of oxygen
(B) Energy of sublimation of magnesium
(C) Lattice energy
(D) First and second ionization energies for magnesium
(E) Electron affinity of O(*g*)

Ⓐ Ⓑ Ⓒ Ⓓ Ⓔ

36. One allotropic form of an element Q is a colorless, crystalline solid at standard conditions. The reaction of Q with an excess of oxygen produces a substance that is a colorless gas at standard conditions. This gaseous product dissolves in cold water to yield a weakly acid solution. The element Q is

(A) boron
(B) carbon
(C) silicon
(D) phosphorus
(E) sulfur

Ⓐ Ⓑ Ⓒ Ⓓ Ⓔ

37. Which of the following species is the strongest Brönsted-Lowry base in water?

(A) NH_3 (B) F^- (C) NH_2^- (D) CO_3^{2-} (E) OH^-

Ⓐ Ⓑ Ⓒ Ⓓ Ⓔ

38. N_2F_2 exists in *cis* and *trans* forms. A bonding model that can account for this observation describes the bond between the two nitrogen atoms as composed of

(A) one σ bond only
(B) one π bond only
(C) two σ bonds
(D) one σ bond and one π bond
(E) one σ bond and two π bonds

Ⓐ Ⓑ Ⓒ Ⓓ Ⓔ

39. Which of the following statements concerning diborane, B_2H_6, is NOT correct?

(A) It is an electron deficient molecule.
(B) There is free rotation about the B–B bond.
(C) The bonding of two hydrogens is of one type whereas the bonding of the other four is of another type.
(D) It forms ammonia addition compounds.
(E) Its final hydrolysis products are hydrogen and boric acid.

Ⓐ Ⓑ Ⓒ Ⓓ Ⓔ

40. (a) $Na^+ + e^- \rightarrow Na$ $\quad E°$ reduction $= -2.71$ V

(b) $2\ H_2O + 2e^- \rightarrow H_2(g) + 2\ OH^-$ $\quad E°$ reduction $= -0.83$ V

Two reactions that may occur at the cathode during the electrolysis of aqueous NaCl are indicated above. The first reaction occurs if the cathode is mercury; the second reaction occurs if the cathode is iron. Part of the accepted explanation for these observations includes the fact that

(A) sodium dissolves in iron and is removed before it has time to react
(B) the reduction potential indicates the second reaction is more probable
(C) sodium and iron form a liquid alloy
(D) iron catalyzes the reaction of sodium with water
(E) the overvoltage for hydrogen evolution on mercury can be in excess of 1.5 volts

Ⓐ Ⓑ Ⓒ Ⓓ Ⓔ

41. The isotope ${}^{222}_{86}Rn$ disintegrates by a series of α and β^- emissions to form ${}^{206}_{82}Pb$. The numbers of α and β^- particles lost are, respectively,

(A) 8 and 4 (B) 6 and 2 (C) 4 and 8 (D) 4 and 4 (E) 2 and 6

Ⓐ Ⓑ Ⓒ Ⓓ Ⓔ

42. Which of the following is the correct distribution of electrons in the $3d$ levels for the $[Co(NH_3)_6]^{3+}$ ion in the ground state? (Atomic number: Co = 27)

(A) 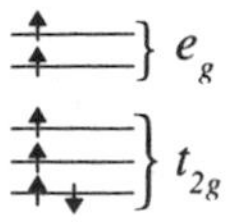 e_g, t_{2g} (B) 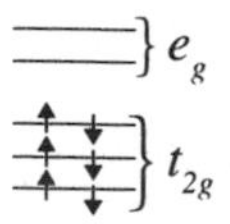 e_g, t_{2g} (C) 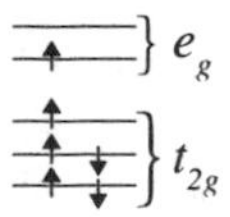 e_g, t_{2g}

(D) 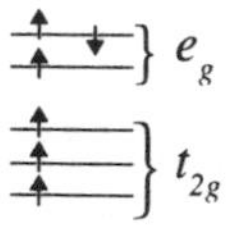 e_g, t_{2g} (E) e_g, t_{2g}

Ⓐ Ⓑ Ⓒ Ⓓ Ⓔ

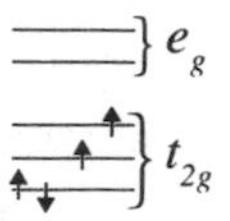

43. Which of the following statements concerning sodium and fluorine is correct? (Atomic numbers: F = 9, Na = 11)

(A) The atomic radius of Na is greater than the atomic radius of F.
(B) The ionic radius of Na^+ is greater than the ionic radius of F^-.
(C) The first ionization energy of Na(g) is greater than the first ionization energy of F(g).
(D) The electron affinity of Na(g) is greater than the electron affinity of F(g).
(E) The number of electrons per Na^+ is greater than the number of electrons per F^-.

Ⓐ Ⓑ Ⓒ Ⓓ Ⓔ

44. According to the molecular orbital theory, the bond order in O_2^- is

(A) $\frac{1}{2}$ (B) 1 (C) $1\frac{1}{2}$ (D) 2 (E) $2\frac{1}{2}$

Ⓐ Ⓑ Ⓒ Ⓓ Ⓔ

45. Which of the following equations represents the most reasonable method of preparing hydrogen persulfide, H_2S_2?

(A) $Na_2S_2 + 2\,H^+ \rightarrow H_2S_2 + 2\,Na^+$
(B) $4\,H_2 + S_8 \rightarrow 4\,H_2S_2$
(C) $H_2O + 2\,H_2S \rightarrow H_2S_2 + 2\,H_2 + 1/2\,O_2$
(D) $2\,H^+ + 2\,S^{2-} \rightarrow H_2S_2 + 2e^-$
(E) $NaOH + 2\,H_2S \rightarrow H_2S_2 + NaH + H_2O$

Ⓐ Ⓑ Ⓒ Ⓓ Ⓔ

46. Which of the following species should be linear?

(A) OF_2 (B) SCl_2 (C) O_3 (D) ICl_2^- (E) NO_2^-

Ⓐ Ⓑ Ⓒ Ⓓ Ⓔ

47. The compound $K_3[FeF_6]$ has a magnetic moment of 5.9 Bohr magnetons, whereas $K_3\ [Fe(CN)_6]$ has a magnetic moment of 2.4 Bohr magnetons. The accepted explanation for this difference includes which of the following statements? (Atomic number: Fe = 26)

(A) Iron has a different oxidation number in the two compounds.
(B) Cyanide ions cause more *d*-orbital splitting than fluoride ions do.
(C) Fluorine is more electronegative than is either carbon or nitrogen.
(D) Cyanide ion is a poor electron donor.
(E) There are fewer unpaired electrons in $K_3[FeF_6]$.

Ⓐ Ⓑ Ⓒ Ⓓ Ⓔ

48. The species in which of the following pairs have decidedly different geometries?

(A) CH_4 and NH_4^+
(B) OF_2 and H_2S
(C) H_3O^+ and PH_3
(D) XeF_4 and BrF_4^-
(E) CF_4 and SF_4

Ⓐ Ⓑ Ⓒ Ⓓ Ⓔ

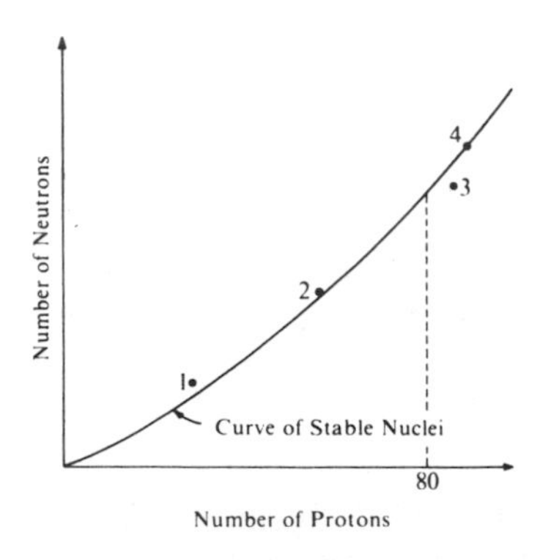

49. On the curve above, certain unstable nuclei are represented by the numbered points. All of the following changes increase the stability of these nuclei EXCEPT:

(A) A nucleus represented by 1 may emit a positron.
(B) A nucleus represented by 2 may emit a gamma ray.
(C) A nucleus represented by 3 may emit a positron.
(D) A nucleus represented by 3 may capture a *K* electron.
(E) A nucleus represented by 4 may emit an alpha particle.

Ⓐ Ⓑ Ⓒ Ⓓ Ⓔ

50. A metal that is NOT found in the free state in nature is

(A) mercury (B) silver (C) copper (D) zinc (E) gold

Ⓐ Ⓑ Ⓒ Ⓓ Ⓔ

51. The fact that Hf^{4+} and Zr^{4+} have almost identical ionic radii has been attributed to the

(A) inert pair effect
(B) Jahn-Teller effect
(C) nephelauxetic effect
(D) chelate effect
(E) lanthanide contraction

Ⓐ Ⓑ Ⓒ Ⓓ Ⓔ

52. All of the following reactions are examples of the Lewis definition of acid-base behavior EXCEPT

(A) $Ni + 4\ CO \rightarrow Ni(CO)_4$
(B) $Ag^+ + 2\ NH_3 \rightarrow Ag(NH_3)_2{}^+$
(C) $BF_3 + NH_3 \rightarrow H_3NBF_3$
(D) $Cl^- + AlCl_3 \rightarrow AlCl_4{}^-$
(E) $Zn + 2\ H^+ \rightarrow Zn^{2+} + H_2$

Ⓐ Ⓑ Ⓒ Ⓓ Ⓔ

53. Compound I + $H_2C_2O_4 \rightarrow [Pt(NH_3)_2C_2O_4]$

Compound II + $H_2C_2O_4 \rightarrow [Pt(NH_3)_2(HC_2O_4)_2]$

$[Pt(NH_3)_2Cl_2]$ is found to exist as two isomers designated as I and II, which are found to react with oxalic acid as indicated above. From this information, one is able to conclude that

(A) I is tetrahedral and II is square planar
(B) I and II are both tetrahedral and I has the *trans* configuration
(C) I and II are both tetrahedral and I has the *cis* configuration
(D) I and II are both square planar and I has the *trans* configuration
(E) I and II are both square planar and I has the *cis* configuration

Ⓐ Ⓑ Ⓒ Ⓓ Ⓔ

54. Which of the following species is paramagnetic? (Atomic numbers: Ti = 22, Cr = 24, Co = 27, Zn = 30, Pd = 46)

(A) $[Zn(NH_3)_4]^{2+}$
(B) $[Co(NH_3)_6]^{3+}$
(C) TiF_4
(D) $[PdCl_4]^{2-}$
(E) $[Cr(NH_3)_6]^{3+}$

Ⓐ Ⓑ Ⓒ Ⓓ Ⓔ

55. The energy sublevel with which of the following quantum-number descriptions can contain the most electrons?

(A) $n = 2, \ell = 1$
(B) $n = 3, \ell = 2$
(C) $n = 4, \ell = 3$
(D) $n = 5, \ell = 0$
(E) $n = 5, \ell = 3, m_\ell = +1$

Ⓐ Ⓑ Ⓒ Ⓓ Ⓔ

56. Which of the following substances does NOT have a metal-metal bond?

(A) $K_2Re_2Cl_8$
(B) $Mn_2(CO)_{10}$
(C) $Mn(CO)_5SnCl_3$
(D) Al_2Cl_6
(E) Hg_2Cl_2

Ⓐ Ⓑ Ⓒ Ⓓ Ⓔ

57. Even though the compounds of the halogen astatine are nearly unknown, which of the following would represent the formula for astatic acid?

(A) HAt (B) HAtO (C) $HAtO_2$ (D) $HAtO_3$ (E) $HAtO_4$

Ⓐ Ⓑ Ⓒ Ⓓ Ⓔ

III. Organic Chemistry

Directions: Each of the questions or incomplete statements below is followed by five suggested answers or completions. Select the one that is best in each case.

58. The structural formulas indicate that one should predict that which of the following compounds has the largest dipole moment?

(A) CCl_4
(B) $O{=}C{=}O$
(C) $(CH_3)_2C{=}C(CH_3)_2$
(D) *trans* $ClCH{=}CHCl$
(E) *cis* $ClCH{=}CHCl$

Ⓐ Ⓑ Ⓒ Ⓓ Ⓔ

59. Which of the following compounds has a bond formed by overlap of $sp - sp^3$ hybrid orbitals?

(A) $CH_3{-}C{\equiv}C{-}H$
(B) $CH_3CH{=}CHCH_3$
(C) $H{-}C{\equiv}C{-}H$
(D) $CH_3CH_2CH_2CH_3$
(E) $CH_2{=}CH{-}CH{=}CH_2$

Ⓐ Ⓑ Ⓒ Ⓓ Ⓔ

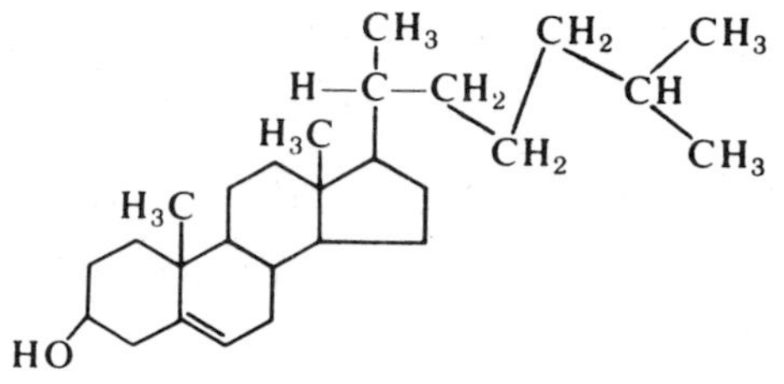

CH3 CH2 CH3
H—C—CH2 CH
H3C CH2 CH3
H3C
HO

60. The formula above represents a member of the class of compounds known as

(A) terpenes (B) alkaloids (C) carbohydrates (D) steroids (E) vitamins

Ⓐ Ⓑ Ⓒ Ⓓ Ⓔ

61. Isomers in which of the following pairs would probably be most nearly equal in stability?

(A) 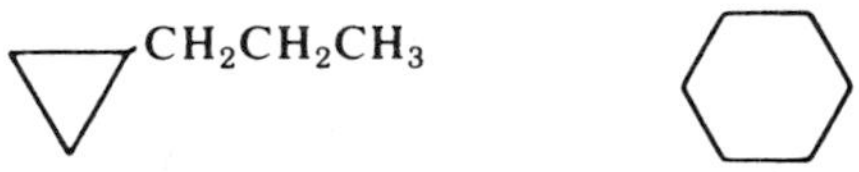 CH2CH2CH3

(B) 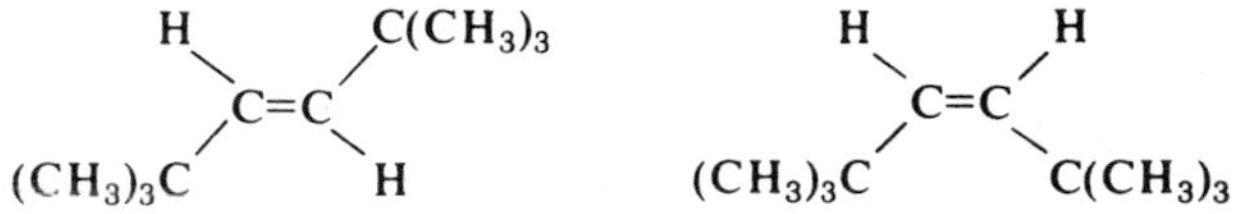 H C(CH3)3 C=C (CH3)3C H   H H C=C (CH3)3C C(CH3)3

(C) CH3 CH3   CH3

(D) CH=CH2   CH3

(E) 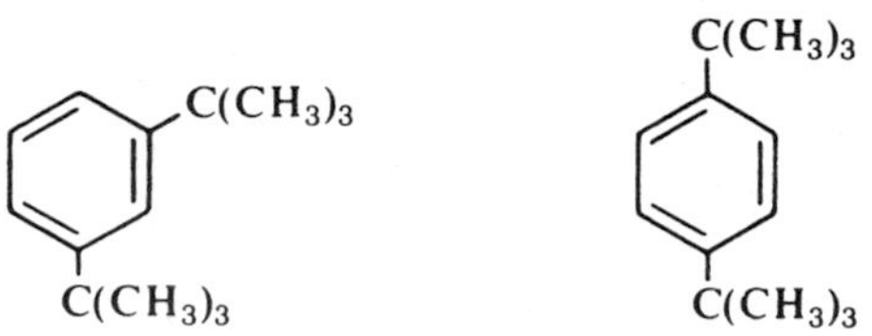 C(CH3)3 C(CH3)3   C(CH3)3 C(CH3)3

Ⓐ Ⓑ Ⓒ Ⓓ Ⓔ

62. Which of the following is the major organic product when acetone, CH_3COCH_3, is allowed to react with hydrogen cyanide in the presence of a catalytic amount of sodium cyanide?

(A) 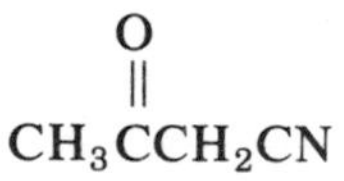 $CH_3\overset{\overset{\displaystyle O}{\|}}{C}CH_2CN$

(B) 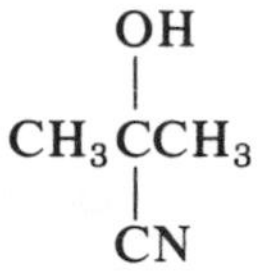 $CH_3\underset{\underset{\displaystyle CN}{|}}{\overset{\overset{\displaystyle OH}{|}}{C}}CH_3$

(C) $(CH_3)_2C{=}C{=}NH$

(D) 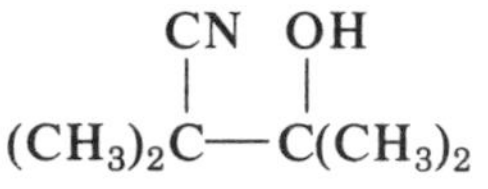 $(CH_3)_2\overset{\overset{\displaystyle CN}{|}}{C}-\overset{\overset{\displaystyle OH}{|}}{C}(CH_3)_2$

(E) 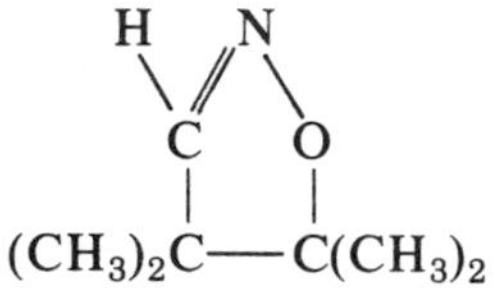 $(CH_3)_2C-C(CH_3)_2$, where the two carbons of the C–C bond are joined in a ring: the left carbon is bonded to a C(H), which is double-bonded to N; N is bonded to O, and O is bonded to the right carbon.

Ⓐ Ⓑ Ⓒ Ⓓ Ⓔ

63. Which of the following compounds has only a single sharp peak in its ^1H nuclear magnetic resonance spectrum?

(A) 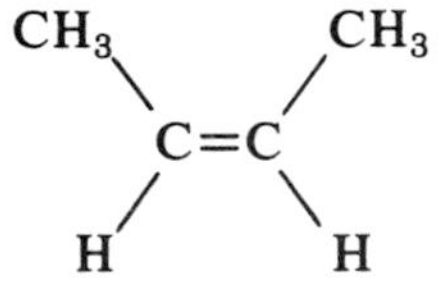 $\begin{matrix} CH_3 & & CH_3 \\ & C{=}C & \\ H & & H \end{matrix}$

(B) 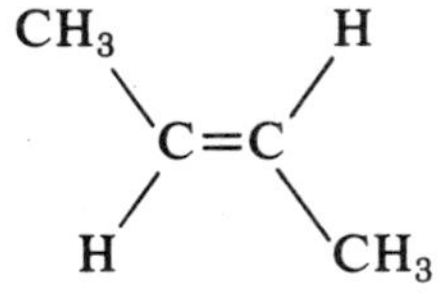 $\begin{matrix} CH_3 & & H \\ & C{=}C & \\ H & & CH_3 \end{matrix}$

(C) $CH_3CH_2CH_2CH_3$

(D) FCH_2CH_2F

(E) $ClCH_2CH_2Cl$

Ⓐ Ⓑ Ⓒ Ⓓ Ⓔ

64. Which of the following is the major organic product of the rapid reaction of cyclohexene with bromine in the dark at 10° C ?

(A) 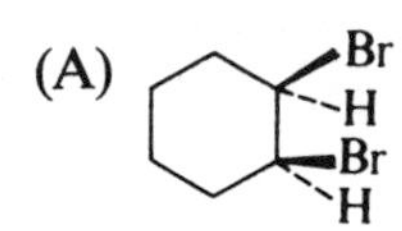

(B) 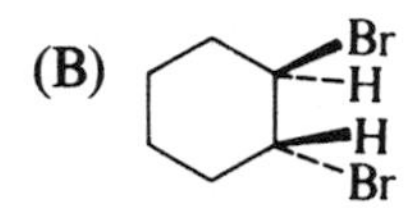

(C) 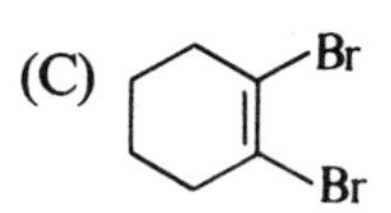

(D) 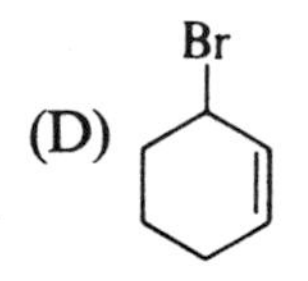

(E) 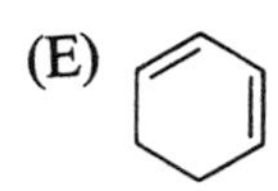

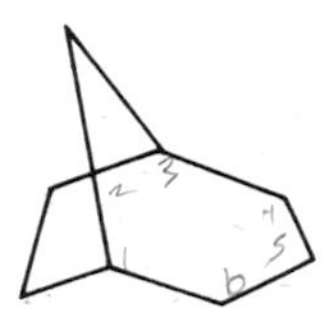

65. The hydrocarbon shown above is

(A) bicyclo[1.1.0]butane

(B) benzyne

(C) bicyclo[3.2.1]octane

(D) bicyclo[2.2.2]octane

(E) tricyclo[3.3.1.1^{3,7}]decane

Ⓐ Ⓑ Ⓒ Ⓓ Ⓔ

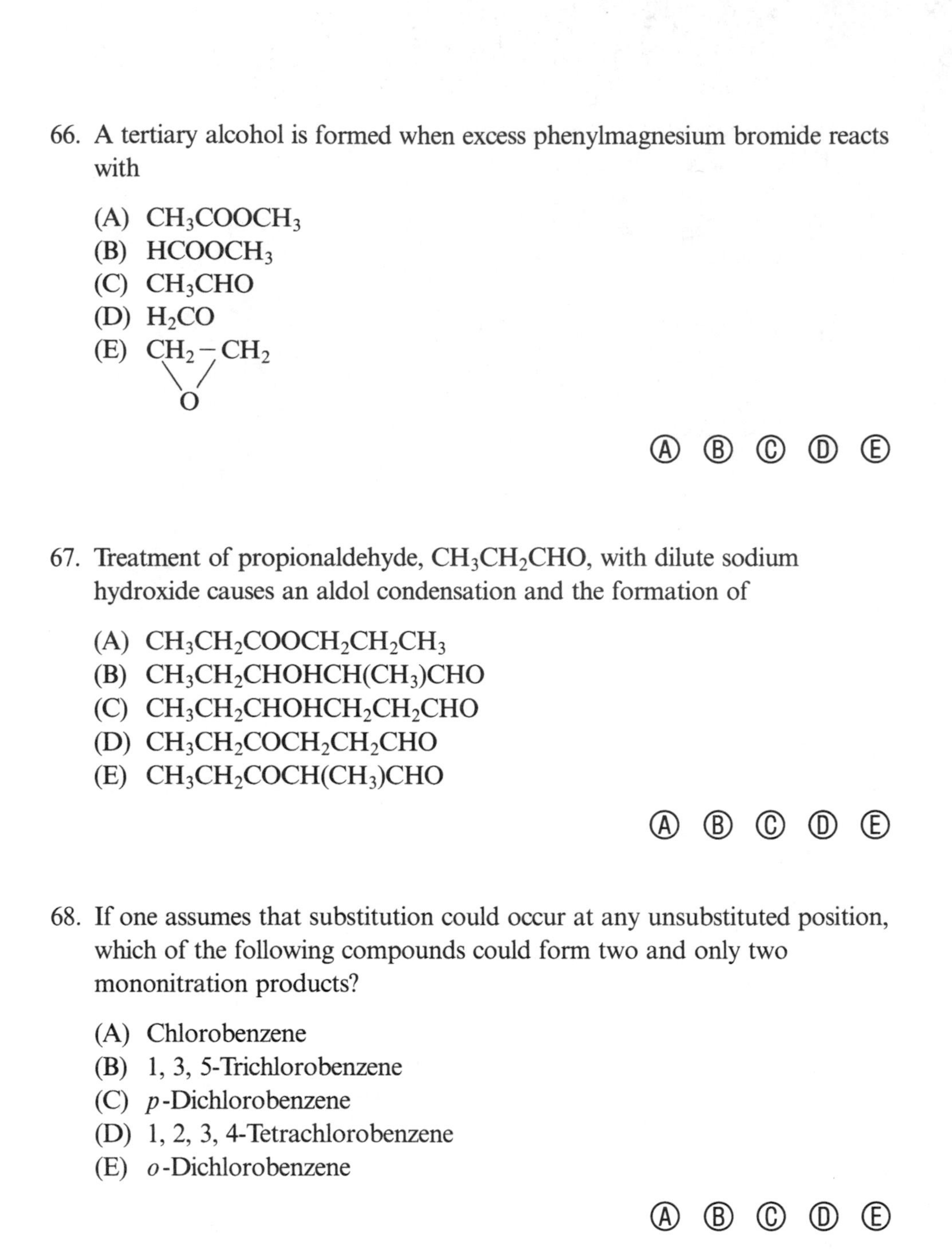

66. A tertiary alcohol is formed when excess phenylmagnesium bromide reacts with

(A) CH_3COOCH_3
(B) $HCOOCH_3$
(C) CH_3CHO
(D) H_2CO
(E) CH_2-CH_2 (ring closed through O)

Ⓐ Ⓑ Ⓒ Ⓓ Ⓔ

67. Treatment of propionaldehyde, CH_3CH_2CHO, with dilute sodium hydroxide causes an aldol condensation and the formation of

(A) $CH_3CH_2COOCH_2CH_2CH_3$
(B) $CH_3CH_2CHOHCH(CH_3)CHO$
(C) $CH_3CH_2CHOHCH_2CH_2CHO$
(D) $CH_3CH_2COCH_2CH_2CHO$
(E) $CH_3CH_2COCH(CH_3)CHO$

Ⓐ Ⓑ Ⓒ Ⓓ Ⓔ

68. If one assumes that substitution could occur at any unsubstituted position, which of the following compounds could form two and only two mononitration products?

(A) Chlorobenzene
(B) 1, 3, 5-Trichlorobenzene
(C) *p*-Dichlorobenzene
(D) 1, 2, 3, 4-Tetrachlorobenzene
(E) *o*-Dichlorobenzene

Ⓐ Ⓑ Ⓒ Ⓓ Ⓔ

69. Which of the following reactions involves a carbocation intermediate?

(A) $2\ CH_3CHO \xrightarrow{OH^-} CH_3CHOHCH_2CHO$

(B) 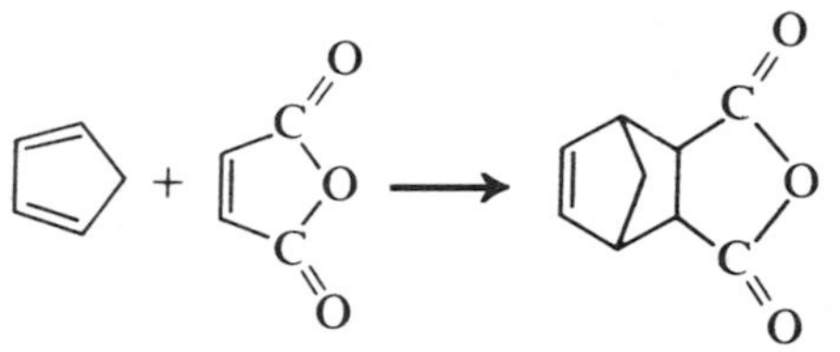 cyclopentadiene + maleic anhydride $\longrightarrow$ bicyclic anhydride adduct

(C) $CH_3CH_3 + Cl_2 \xrightarrow{h\nu} CH_3CH_2Cl + HCl$

(D) $(C_6H_5)_2CHBr + H_2O \longrightarrow (C_6H_5)_2CHOH + HBr$

(E) $CH_3CH{=}CH_2 + CH_2N_2 \xrightarrow{h\nu} CH_3-\underset{\diagdown CH_2 \diagup}{CH—CH_2} + N_2$

Ⓐ Ⓑ Ⓒ Ⓓ Ⓔ

70. The most acidic of the following compounds is

(A) phenol
(B) *p*-aminophenol
(C) *p*-nitrophenol
(D) *m*-nitrophenol
(E) 2, 6-di-*t*-butylphenol

Ⓐ Ⓑ Ⓒ Ⓓ Ⓔ

71. Which of the following molecules is NOT considered to be aromatic?

(A)
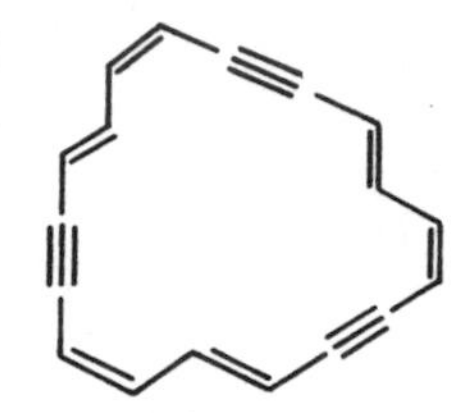

(B)
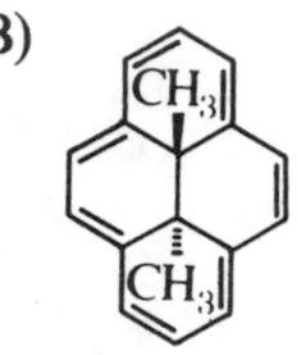

(C)
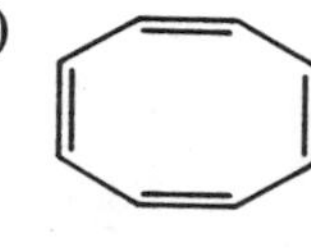

(D)
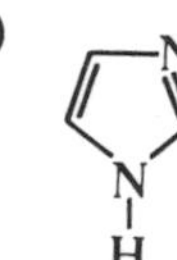

(E)
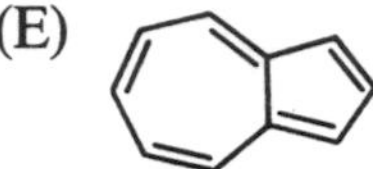

Ⓐ Ⓑ Ⓒ Ⓓ Ⓔ

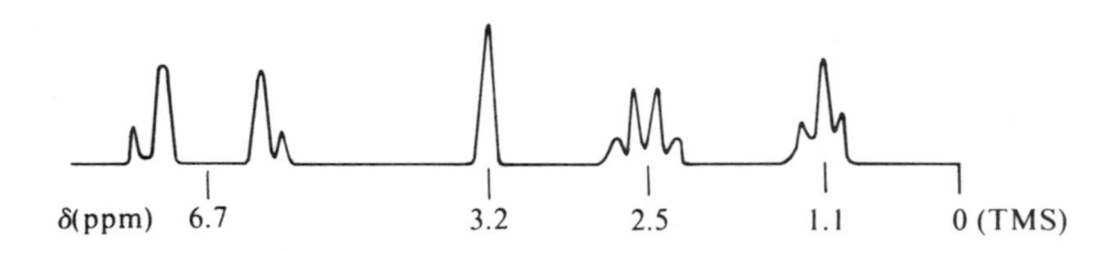

72. A compound with the molecular formula $C_8H_{11}N$ showed infrared absorption near 2.9 microns (3,448 cm^{-1}), ultraviolet absorption at 235 nanometers ($\epsilon = 1{,}480$), and the nuclear magnetic resonance spectrum depicted above. Which of the following structures is consistent with these data?

(A) $N(CH_3)_2$
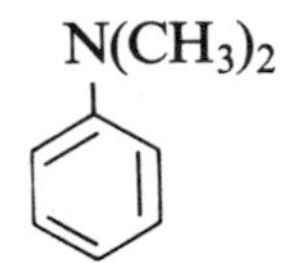

(B) $NHCH_2CH_3$
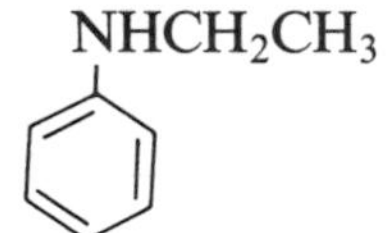

(C) CH_2CH_3
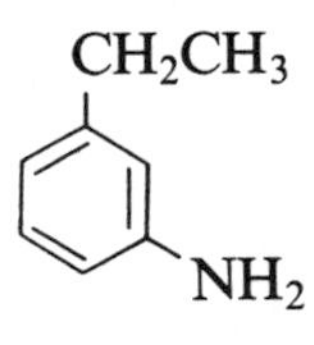
NH_2

(D) CH_2NH_2
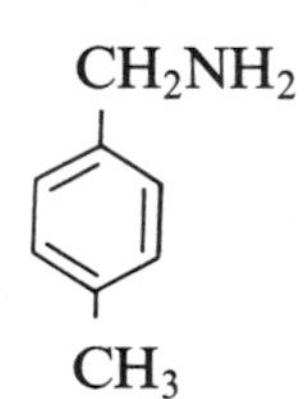
CH_3

(E)
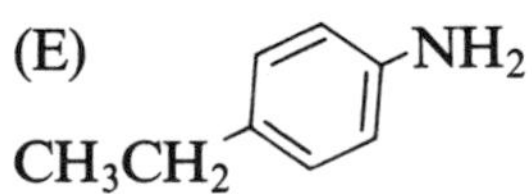

Ⓐ Ⓑ Ⓒ Ⓓ Ⓔ

73.

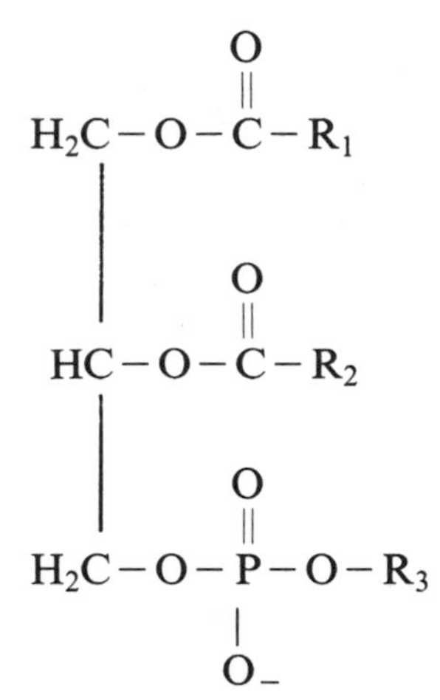

Phospholipids of the general formula shown above are arranged in lipid bilayers visualized as follows:

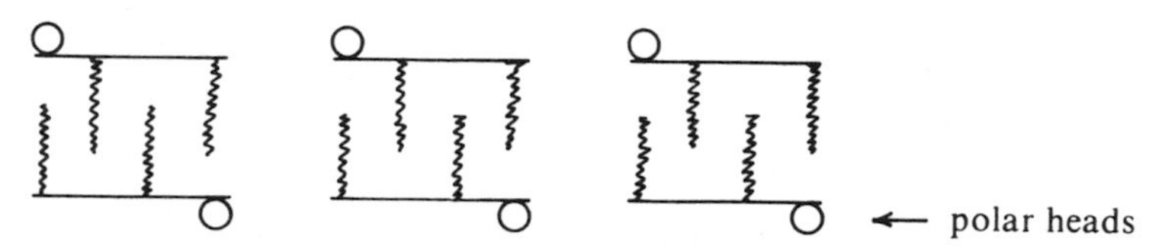

The polar heads which are exposed to an aqueous environment represent the

(A) fatty acid group R_1 only
(B) fatty acid groups R_1 and R_2
(C) phosphate diester function
(D) carbonyl function of the carboxylic acid
(E) glycerol backbone

Ⓐ Ⓑ Ⓒ Ⓓ Ⓔ

74. The preferred conformation of *trans*-1, 4-dimethylcyclohexane has the cyclohexane ring in the

(A) chair form with both methyl groups equatorial
(B) chair form with both methyl groups axial
(C) chair form with one methyl group axial and one equatorial
(D) boat form with the methyl groups pointing toward the center of the ring
(E) boat form with the methyl groups pointing away from the ring

Ⓐ Ⓑ Ⓒ Ⓓ Ⓔ

75. A hexapeptide is hydrolyzed to the dipeptides Ileu-Val, Ala-Pro, and Lys-Leu. Carboxypeptidase acts on the hexapeptide to liberate valine, and 2, 4-dinitrofluorobenzene reacts with the hexapeptide to yield, after hydrolysis, 2, 4-dinitrophenylalanine. Which of the following is the amino acid sequence of the hexapeptide?

(A) Ala-Pro-Lys-Leu-Ileu-Val
(B) Val-Ileu-Lys-Leu-Pro-Ala
(C) Ileu-Val-Ala-Pro-Lys-Leu
(D) Val-Ala-Pro-Lys-Leu-Ileu
(E) Lys-Leu-Ala-Pro-Ileu-Val

76. Of the following, which compound is the strongest Brönsted-Lowry acid?

(A) $CH_3-CH_2-CH_2-OH$
(B) $CH_3-CHBr-COOH$
(C) $CH_3-C\equiv C-H$
(D) CH_3-CH_2-COOH
(E) $CH_3-CH_2-SO_3H$

77. Which of the following reactions is NOT a typical reaction of the carbonyl group ($>C=O$) ?

(A) $>C=O + H_2 \xrightarrow{Pt} >CHOH$

(B) $>C=O + H_2N-OH \longrightarrow >C=N-OH + H_2O$

(C) $>C=O + CH_3Br \longrightarrow >C(Br)-O-CH_3$

(D) $>C=O + HCN \xrightarrow{OH^-} >C(CN)-OH$

(E) $>C=O + RMgBr \longrightarrow >C(R)-OMgBr$

Ⓐ Ⓑ Ⓒ Ⓓ Ⓔ

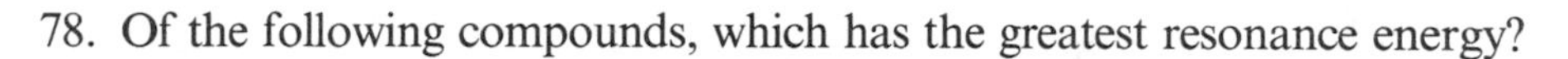
78. Of the following compounds, which has the greatest resonance energy?

(A) 2-Butene
(B) Cyclohexene
(C) 1,3-Butadiene
(D) 2,3-Butanedione

(E) 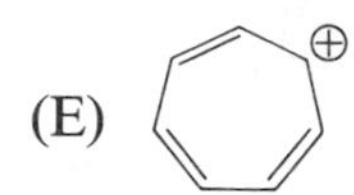

79. Which of the following compounds should be expected to have the highest boiling point?

(A) $CH_3CH_2OCH_2CH_3$
(B) $CH_3CH_2COCH_3$
(C) $CH_3CH_2CH_2CH_2OH$
(D) $CH_3CH_2CH_2COOH$
(E) $CH_3CH_2CH_2CH_2CH_3$

80. The conversion of 1-butene to 1-butanol is accomplished in synthetically useful yield by treating the 1-butene with

(A) HCl followed by H_2O
(B) B_2H_6 followed by H_2O_2 and NaOH

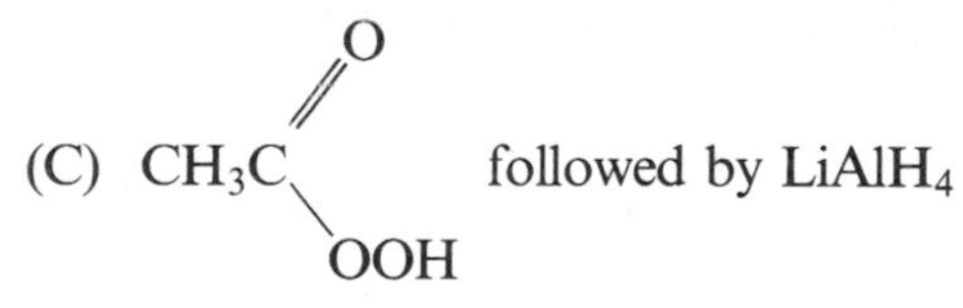
(C) $CH_3C(=O)OOH$ followed by $LiAlH_4$

(D) $KMnO_4$
(E) H_2(Pt) in H_2O

81.

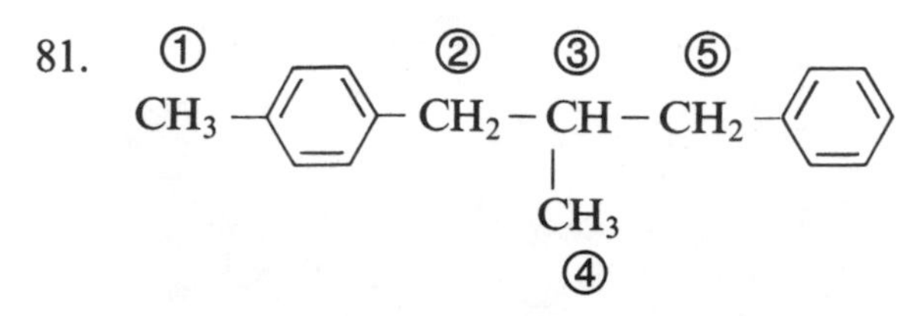

In the structural formula shown above, which of the numbered hydrogen atoms is LEAST susceptible to substitution by Cl radicals? (Assume that the reaction occurs by a free radical mechanism under reaction conditions in which substitution of hydrogen is selective.)

(A) 1 (B) 2 (C) 3 (D) 4 (E) 5

Ⓐ Ⓑ Ⓒ Ⓓ Ⓔ

82. Which of the following compounds would be most susceptible to electrophilic attack by nitronium ions?

(A)

(B) NO_2 CF_3

(C) CH_3 NO_2

(D) CH_3

(E) OCH_3 CH_3

Ⓐ Ⓑ Ⓒ Ⓓ Ⓔ

83. The concept that best explains the greater volatility of *o*-nitrophenol over *p*-nitrophenol during steam distillation of a mixture of the two compounds is

(A) hyperconjugation
(B) hydrogen bonding
(C) the *ortho*-effect
(D) resonance
(E) symmetry

Ⓐ Ⓑ Ⓒ Ⓓ Ⓔ

84. Which of the following carbocations would most easily undergo a 1,2-hydride shift?

(A) $(CH_3)_3C^{\oplus}$

(B) $C_6H_5\overset{\oplus}{C}HCH_3$

(C) $CH_3\overset{\oplus}{C}HC(CH_3)_3$

(D) $CH_3\overset{\oplus}{C}HCH(CH_3)_2$

(E) $(CH_3)_2\overset{\oplus}{C}CH_2CH_3$

Ⓐ Ⓑ Ⓒ Ⓓ Ⓔ

85. Which of the following compounds undergoes dehydrohalogenation most rapidly in boiling ethanol by an E_1 mechanism?

(A) $CH_3CBr=C(CH_3)_2$

(B) $CH_3CHBrCH(CH_3)_2$

(C) $C_6H_5-CHBrCH_3$

(D) $O_2N-C_6H_4-CHBrCH_3$

(E) $CH_3O-C_6H_4-CHBrCH_3$

Ⓐ Ⓑ Ⓒ Ⓓ Ⓔ

86. The reduction of cyclohexanone in isopropyl alcohol in the presence of aluminum isopropoxide can be used to prepare cyclohexanol in a good yield only if which of the following conditions is fulfilled?

(A) The reaction is carried out at 0° C.
(B) The reaction is irradiated with sunlight.
(C) The acetone formed in the reaction is distilled away as the reaction is taking place.
(D) The cyclohexanol is distilled away from the reaction mixture as the reaction is taking place.
(E) The reaction is catalyzed with mercuric chloride.

Ⓐ Ⓑ Ⓒ Ⓓ Ⓔ

87. A carboxylic acid is formed in high yield by which of the following reactions?

(A) 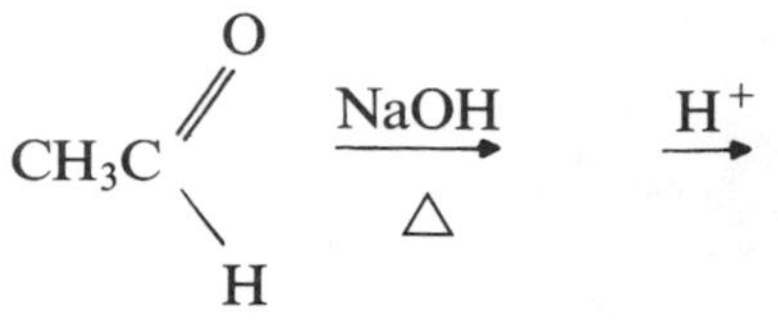 $CH_3C(=O)H \xrightarrow[\Delta]{NaOH} \xrightarrow{H^+}$

(B)  (cyclobutane ring bearing H and CCl_2CH_3) $\xrightarrow[H^+]{H_2O}$

(C) $CH_3C{\equiv}CCH_3 \xrightarrow[Hg^{2+}]{H_2SO_4,\ H_2O}$

(D) $CH_3CH_2\overset{O}{\overset{\|}{C}}CCl_3 \xrightarrow{NaOH} \xrightarrow{H^+}$

(E) 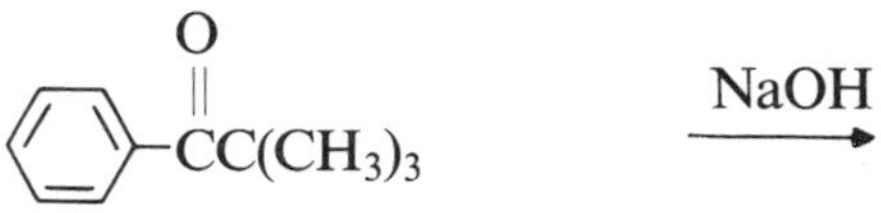 $C_6H_5\overset{O}{\overset{\|}{C}}C(CH_3)_3 \xrightarrow{NaOH}$

Ⓐ Ⓑ Ⓒ Ⓓ Ⓔ

88. Which of the following is the predominant product in the reaction of HOBr with propene?

(A) 2-Bromo-1-propanol
(B) 3-Bromo-1-propanol
(C) 2-Bromo-2-propanol
(D) 1-Bromo-2-propanol
(E) 1, 2-Dibromopropane

Ⓐ Ⓑ Ⓒ Ⓓ Ⓔ

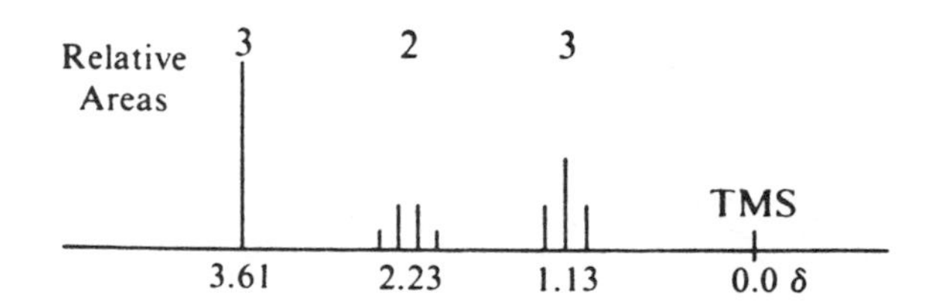

89. Which of the following compounds has the ^1H nuclear magnetic resonance spectrum indicated above?

(A) $CH_3-\overset{O}{\overset{\|}{C}}-CH_2-OCH_3$

(B) $CH_3-CH_2-\overset{O}{\overset{\|}{C}}-OCH_3$

(C) $CH_3-CH_2-\overset{O}{\overset{\|}{C}}-CH_3$

(D) $CH_3-CH_2-O-CH_3$

(E) $CH_3-\overset{O}{\overset{\|}{C}}-O-CH_2-CH_3$

Ⓐ Ⓑ Ⓒ Ⓓ Ⓔ

90. Which of the following is NOT an *ortho*/*para*-directing group in electrophilic aromatic substitution?

(A) NH_2 (B) OCH_3 (C) CH_3 (D) CF_3 (E) F

Ⓐ Ⓑ Ⓒ Ⓓ Ⓔ

91. Which of the following generally reacts with ketones to form olefins?

(A) $(C_6H_5-)_3P-CH_2^-$

(B) $NaOC_2H_5$

(C) CH_3MgBr

(D) $C_6H_5-N=C=O$

(E) $C_6H_{11}-N=C=N-C_6H_{11}$

Ⓐ Ⓑ Ⓒ Ⓓ Ⓔ

92. The dimer formed by heating 1,3-butadiene has which of the following structures?

(A)

(B)

(C) 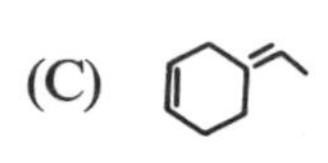

(D) 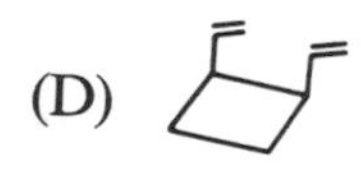

(E) 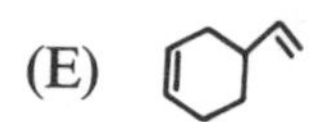

93. Which of the following reactions involves an enolate anion intermediate?

(A) Alkylation of an enamine with methyl iodide
(B) Self-condensation of acetaldehyde catalyzed by sodium hydroxide
(C) Solvolysis of *t*-butyl chloride in aqueous acetone
(D) Nucleophilic displacement of iodide from 2,4,6-trinitroiodobenzene by ethoxide ion
(E) Nitration of toluene

94. A laboratory preparation of pure propane involves

(A) the oxidation of propionic acid
(B) the action of sodium on propyl bromide
(C) the action of water on *n*-propyl magnesium bromide
(D) treating *n*-propyl alcohol with zinc and hydrochloric acid
(E) heating calcium propionate

95. The product of the reaction of sodium phenoxide with sodium chloroacetate is acidified with hydrochloric acid and extracted with ether. The ether contains phenoxyacetic acid and a small amount of phenol. Separation of these two compounds can be accomplished most easily by

(A) extracting the ether solution with aqueous sodium hydroxide and then acidifying the aqueous layer
(B) extracting the ether solution with aqueous sodium bicarbonate and then acidifying the aqueous layer
(C) removing the ether and then crystallizing the product
(D) removing the ether and then distilling the product
(E) removing the ether and then steam-distilling the product

Ⓐ Ⓑ Ⓒ Ⓓ Ⓔ

96. Which of the following halides generally undergoes bimolecular nucleophilic substitution (S_N2) reactions most readily?

(A) $CH_3\text{—}CH(CH_3)\text{—}Br$

(B) C_6H_5-Br

(C) 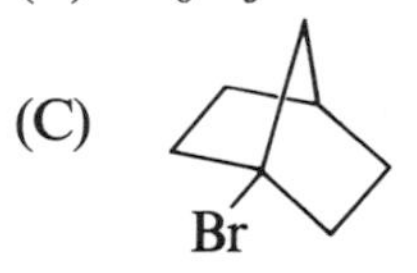

(D) 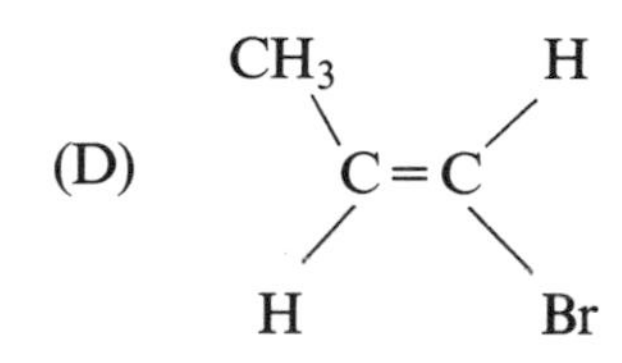

(E) $C_6H_5-CH_2-Br$

Ⓐ Ⓑ Ⓒ Ⓓ Ⓔ

97. Which of the following compounds does NOT show tautomerism?

(A)

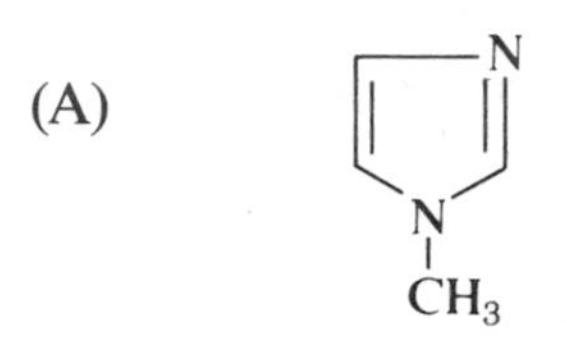

(B)

N—N
N
H
CH$_3$

(C)

O O
C C
CH$_3$ CH CH$_3$
CH$_3$

(D)

O O
C C
CF$_3$ CH$_2$ CH$_3$

(E)

CH$_3$ N
N
H

98. An unknown compound was found to react with sodium hydride with the evolution of hydrogen. The compound could not be acetylated under the normal conditions for acetylation. The compound resisted oxidation under mild conditions and under vigorous conditions yielded only products of molecular weight much smaller than the unknown. On the basis of these facts the compound is

(A) a primary alcohol
(B) an aldehyde
(C) a secondary alcohol
(D) a tertiary alcohol
(E) a secondary amine

99. Which of the following reactions does NOT represent a practical method for nucleophilic displacement of a halide ion from an organic halide?

(A) $CH_3CH_2CH_2CH_2Br + CN^- \longrightarrow CH_3CH_2CH_2CH_2CN + Br^-$

(B) 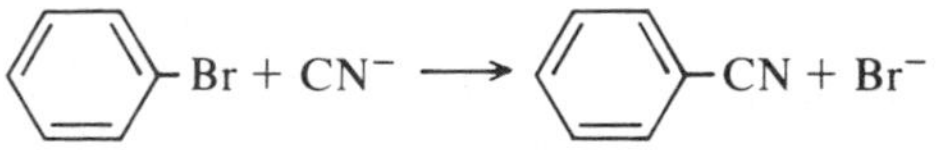 C_6H_5–Br + CN^- ⟶ C_6H_5–CN + Br^-

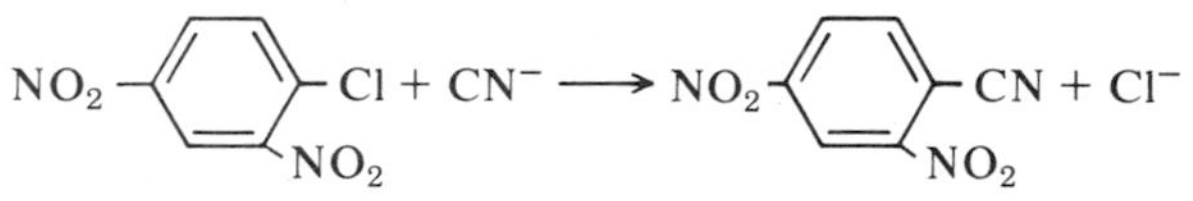

(C) 2,4-$(NO_2)_2C_6H_3$–Cl + CN^- ⟶ 2,4-$(NO_2)_2C_6H_3$–CN + Cl^-

(D) 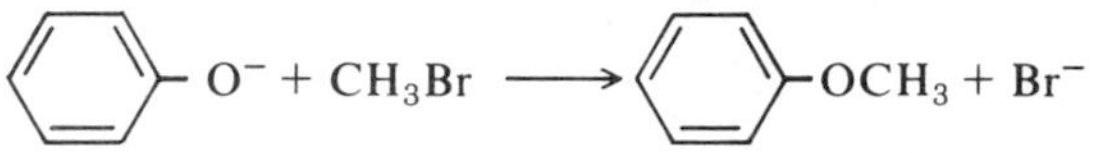 C_6H_5–O^- + CH_3Br ⟶ C_6H_5–OCH_3 + Br^-

(E) 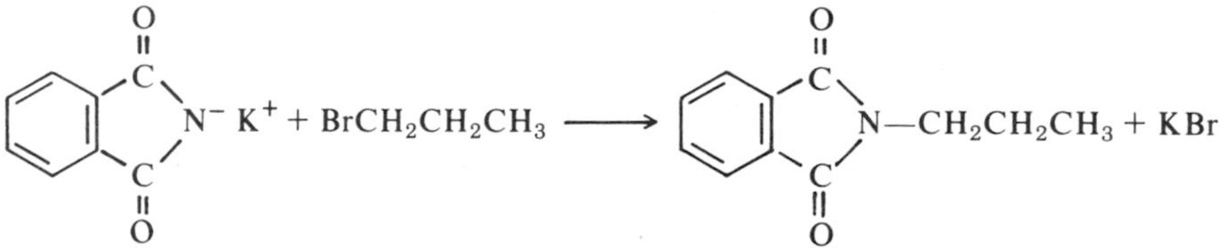 Potassium phthalimide ($N^- K^+$) + $BrCH_2CH_2CH_3$ ⟶ N–$CH_2CH_2CH_3$ phthalimide + KBr

Ⓐ Ⓑ Ⓒ Ⓓ Ⓔ

100. Compounds that are incapable of optical activity include which of the following?

I. 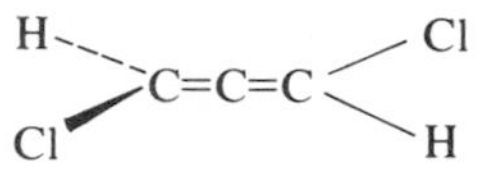 (H)(Cl)C=C=C(Cl)(H)

II. 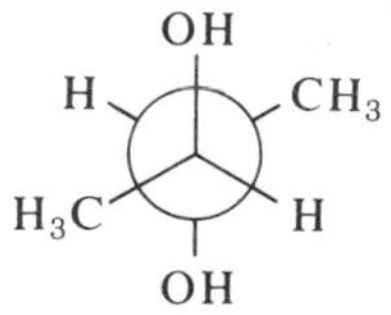 Newman projection: front OH, H, H_3C; back CH$_3$, H, OH

III. 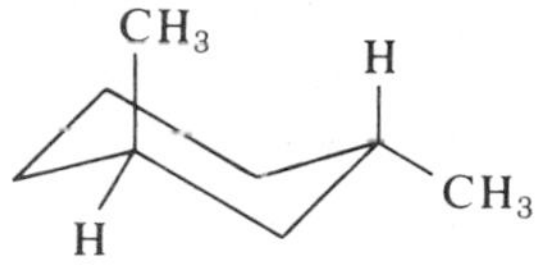 Cyclohexane with CH$_3$ and H on each of two carbons

IV. 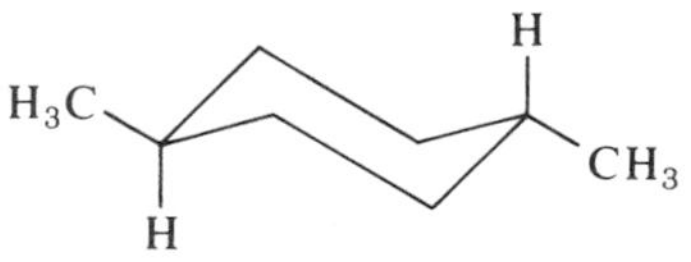 Cyclohexane with H_3C and H, and H and CH_3 on two carbons

(A) I only (B) III only (C) I and II (D) II and IV
(E) III and IV

Ⓐ Ⓑ Ⓒ Ⓓ Ⓔ

101. The addition of traces of benzoyl peroxide at 80° C may be expected to accelerate all of the following reactions EXCEPT

(A) $C_6H_5CH(CH_3)_2 \xrightarrow{O_2} C_6H_5C(OOH)(CH_3)_2$

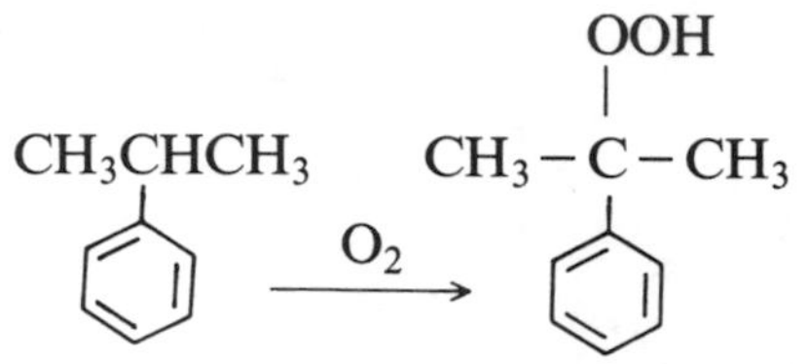

CH$_3$CHCH$_3$ (on benzene ring) $\xrightarrow{O_2}$ CH$_3$–C(OOH)–CH$_3$ (on benzene ring)

(B) $C_6H_5CH_3 \xrightarrow{\text{N-bromosuccinimide (N–Br)}} C_6H_5CH_2Br$

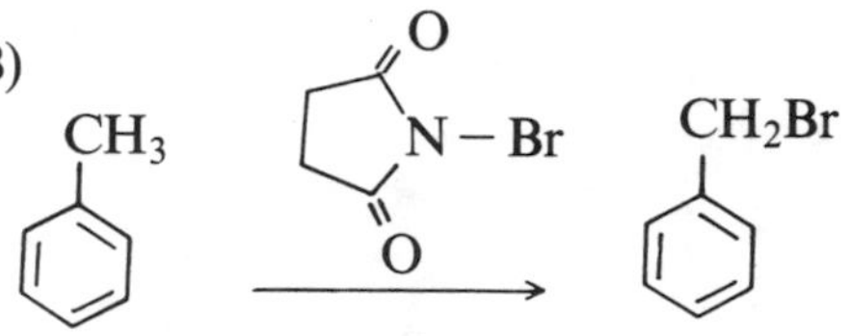

(C) HBr + cyclohexene $\longrightarrow$ bromocyclohexane (Br)

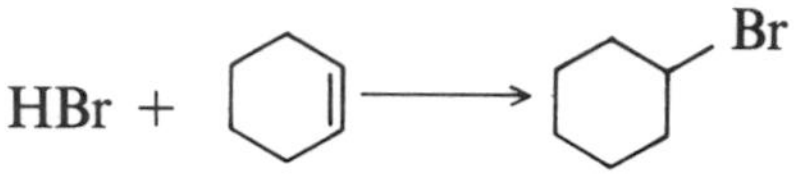

(D) $(CH_3)_3CH + Br_2 \longrightarrow (CH_3)_3CBr + HBr$

(E) $CH_3\overset{O}{\overset{\|}{C}}CH_2CO_2H \longrightarrow CH_3\overset{O}{\overset{\|}{C}}CH_3 + CO_2$

Ⓐ Ⓑ Ⓒ Ⓓ Ⓔ

IV. Physical Chemistry

Directions: Each of the questions or incomplete statements below is followed by five suggested answers or completions. Select the one that is best in each case.

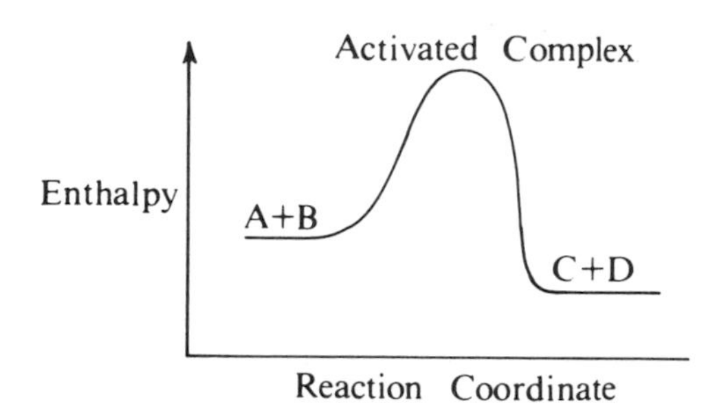

102. Relative values for the enthalpies involved in the reaction A + B → C + D are shown in the diagram above. Because the enthalpy of C + D is lower than that of A + B, one knows that

(A) the reaction is irreversible
(B) a catalyst for the reaction is impossible
(C) the reaction is exothermic
(D) the activation energy required for the reverse reaction is lower than that for the forward reaction
(E) the activated complex for the reverse reaction is a different species from that for the forward reaction

Ⓐ Ⓑ Ⓒ Ⓓ Ⓔ

103. Information regarding the energies required to stretch and bend chemical bonds can be obtained most directly from studies of which of the following regions of the spectrum?

(A) Infrared (B) Visible (C) Ultraviolet (D) X-ray (E) Microwave

Ⓐ Ⓑ Ⓒ Ⓓ Ⓔ

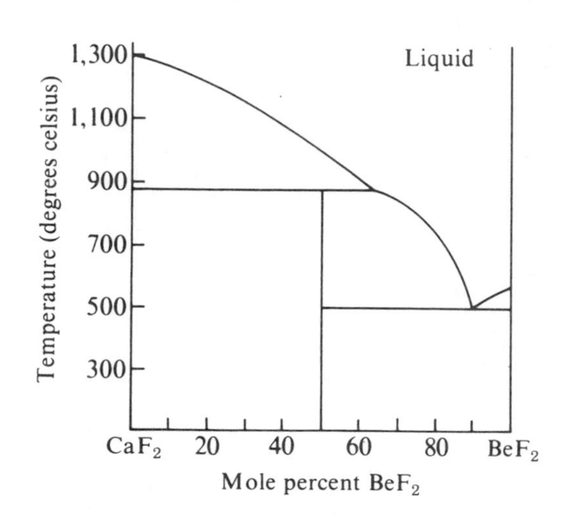

104. The figure above shows a phase diagram for the system CaF_2, BeF_2 at a pressure of 1 atmosphere. If a sample that is 75-mole percent BeF_2 is cooled from 1,100° C to 300° C, the phases that coexist in equilibrium at 300° C are

(A) solid CaF_2 and liquid solution
(B) liquid CaF_2 and solid CaF_2
(C) liquid BeF_2 and solid BeF_2
(D) solid BeF_2 and solid $BeF_2 \cdot CaF_2$
(E) solid BeF_2 and liquid solution

105. Under conditions for which the ideal gas law is applicable, the observed density of a gas was found to be almost twice as great as the value calculated from the usual molecular formula. This finding suggests that

(A) many of the molecules of the gas are dissociated into atoms
(B) correction must be made for the volume occupied by the molecules
(C) many of the molecules of gas undergo association to form dimers
(D) the molecular weight must be less than the theoretical value
(E) the virial equation should be used in the calculation

106. For a certain first-order gas-phase reaction $A \rightarrow B + C$, the time required for half of an initial amount of A to decompose is 8.8 minutes. If the initial pressure of A is 400 millimeters of mercury, the time required to reduce the partial pressure of A to 50 millimeters of mercury is

(A) 1.1 min
(B) 17.6 min
(C) 26.4 min
(D) 35.2 min
(E) 70.4 min

Ⓐ Ⓑ Ⓒ Ⓓ Ⓔ

107. 1. $N_2(g) + 3\ H_2(g) \rightarrow 2\ NH_3(g)$

2. $2\ NaCl(\ell) \rightarrow Na_2Cl_2(g)$

Would the standard entropy change for the reactive systems above be positive (+), negative (−), or near zero (0) ?

	Equation 1	Equation 2
(A)	+	−
(B)	−	+
(C)	0	−
(D)	+	0
(E)	−	0

Ⓐ Ⓑ Ⓒ Ⓓ Ⓔ

108. If, at absolute temperature T, the free energy change of a reaction is ΔG and the enthalpy change is ΔH, then ΔS, the entropy change for the reaction, is equal to

(A) $\dfrac{\Delta G - \Delta H}{T}$

(B) $\dfrac{\Delta H - \Delta G}{T}$

(C) $T(\Delta G - \Delta H)$

(D) $T(\Delta H - \Delta G)$

(E) $\dfrac{T}{\Delta H - \Delta G}$

Ⓐ Ⓑ Ⓒ Ⓓ Ⓔ

109. The energy of which of the following systems is NOT quantized?

(A) A particle in a box
(B) A particle in free space
(C) A rigid rotor
(D) A simple harmonic oscillator
(E) An electron in an atomic orbital

Ⓐ Ⓑ Ⓒ Ⓓ Ⓔ

110. The Joule-Thomson coefficient of a gas, $(\partial T/\partial P)_H$, is measured under conditions of constant

(A) energy (B) enthalpy (C) entropy (D) pressure (E) temperature

Ⓐ Ⓑ Ⓒ Ⓓ Ⓔ

111. Which of the following is characteristic of an adiabatic change?

(A) Heat is absorbed from outside the system.
(B) Heat is given up to the surroundings.
(C) No heat enters or leaves the system.
(D) The temperature remains constant.
(E) The internal energy of the system remains constant.

Ⓐ Ⓑ Ⓒ Ⓓ Ⓔ

112. Which of the following choices correctly describes effects associated with the fact that even ideal solutions exhibit osmotic pressure?

	Change in Entropy When More Solvent Is Added to an Ideal Solution	Change in Vapor Pressure of Solvent When Solute Is Added
(A)	Decrease	Increase
(B)	No change	Increase
(C)	No change	No change
(D)	Increase	Increase
(E)	Increase	Decrease

Ⓐ Ⓑ Ⓒ Ⓓ Ⓔ

113. In an ethanol-water solution in which the mole fraction of water is 0.400, the partial molal volume of water is 16.0 milliliters and that of ethanol is 57.0 milliliters. The volume of 1.00 mole of this solution is

(A) $(0.600)(16.0) + (0.400)(57.0)$ ml

(B) $\frac{(0.600)(16.0) + (0.400)(57.0)}{(0.400)(0.600)}$ ml

(C) $\frac{0.400}{16.0} + \frac{0.600}{57.0}$ ml

(D) $(0.400)(16.0) + (0.600)(57.0)$ ml

(E) $\frac{16.0}{0.600} + \frac{57.0}{0.400}$ ml

Ⓐ Ⓑ Ⓒ Ⓓ Ⓔ

114. The Morse curve of potential energy V versus the interatomic distance d for the H_2 molecule is best represented by which of the following curves?

(A) 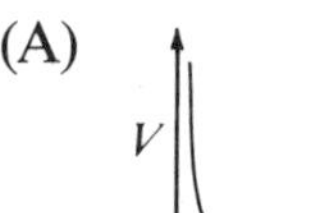 

(B) 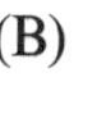 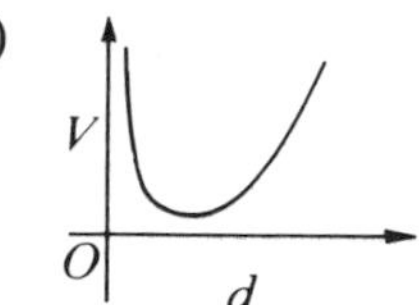

(C) 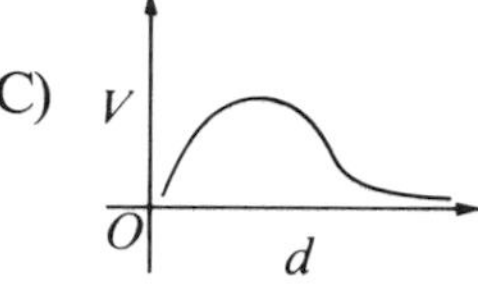

(D) 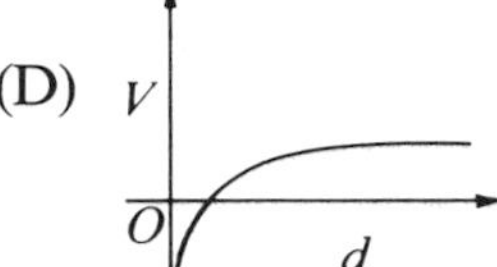

(E)  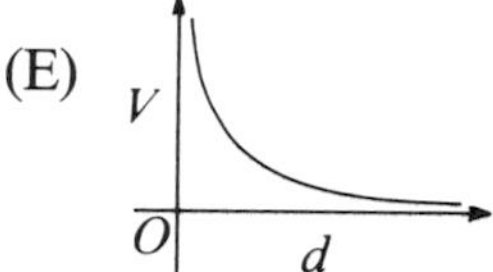

115. If the reversible addition of 0.02 calorie of heat to a sample of a substance at constant temperature increases the entropy of the sample by 0.05 calorie/degree, the temperature of the sample is

(A) 2.5 K
(B) 0.4 K
(C) 0.3 K
(D) 0.001 K
(E) none of the above

116. Which of the following gas properties is NOT directly related to σ, the molecular diameter?

(A) The viscosity coefficient
(B) The mean free path
(C) The number of collisions per second for one molecule
(D) The average speed of a molecule
(E) The parameter b in the van der Waals equation of state

117. The unit cell for metallic silver is a face-centered cube 4.08 Å on a side. What is the atomic radius of silver?

(A) 1.02Å (B) 1.44Å (C) 2.04Å (D) 2.88Å (E) 4.08Å

Ⓐ Ⓑ Ⓒ Ⓓ Ⓔ

118. A detailed study of the infrared spectrum of a diatomic molecule can lead directly to all of the following pieces of information EXCEPT the

(A) translational energy of the molecule
(B) force constant of the bond
(C) relative distribution of molecules among the allowed rotational levels
(D) average bond distance
(E) relative isotopic abundance

Ⓐ Ⓑ Ⓒ Ⓓ Ⓔ

119. $$IO_3^- + 8\,I^- + 6\,H^+ \longrightarrow 3\,I_3^- + 3H_2O$$

A mechanism proposed for the reaction above is as follows:

Step 1: $IO_3^- + 2\,H^+ \overset{K}{\rightleftarrows} H_2IO_3^+$ (fast, to equilibrium)

Step 2: $I^- + H_2IO_3^+ \overset{k_1}{\longrightarrow} I_2O_2 + H_2O$ (slow)

Step 3: $I^- + I_2O_2 \overset{k_2}{\longrightarrow} I_2 + IO_2^-$ (fast)

Step 4: $I_2 + I_3^- \overset{k_3}{\longrightarrow} I_3^-$ (fast)

Subsequent steps fast

Based on this mechanism, the rate law is which of the following? (k' is an appropriate composite of the other relevant constants.)

(A) Rate $= k'[H^+]^2\,[IO_3^-]\,[I^-]$
(B) Rate $= k'[H^+]\,[IO_3^-]\,[I^-]$
(C) Rate $= k'[IO_3^-]\,[I^-]$
(D) Rate $= k'[H^+]^6\,[IO_3^-]\,[I^-]^8$
(E) Rate $= k'[IO_3^-]\,[I^-]^8$

Ⓐ Ⓑ Ⓒ Ⓓ Ⓔ

120. The pH of a 0.1-molar solution of the acid HQ is 3. What is the value for the ionization constant of this acid?

(A) 0.3
(B) 10^{-3}
(C) 10^{-5}
(D) 10^{-7}
(E) 10^{-9}

Ⓐ Ⓑ Ⓒ Ⓓ Ⓔ

121. If s represents an s orbital, which of the following expressions is the best (nonnormalized) wave function for a hydrogen molecule $H_a{-}H_b$?

(A) $s_a(1)s_b(2)$
(B) $s_a(1) + s_b(2)$
(C) $s_a(1)s_b(1) + s_a(2)s_b(2)$
(D) $s_a(1)s_b(2) + s_a(2)s_b(1)$
(E) $s_a(1)s_b(2) - s_a(2)s_b(1)$

Ⓐ Ⓑ Ⓒ Ⓓ Ⓔ

122. If the value of $\Delta G°$ for a given reaction is known, one may determine all of the following EXCEPT the

(A) direction of spontaneous change at standard conditions
(B) position of equilibrium at the temperature for which $\Delta G°$ is known
(C) maximum amount of useful work that can be produced at standard conditions
(D) stability of the products as compared to the reactants at standard conditions
(E) usefulness of a catalyst in controlling the reaction

Ⓐ Ⓑ Ⓒ Ⓓ Ⓔ

123. $N_2(g) + 3\ H_2(g) = 2\ NH_3(g)$ $\quad \Delta H° = -92.0$ kilojoules at 25° C

To calculate ΔH, the change in enthalpy, at 100° C for the reaction above, one needs what additional information?

(A) The equilibrium constant for the reaction at 100° C
(B) The molar heat capacities of the reactants and the products as a function of temperature
(C) $\Delta E°$, the standard internal energy change for the reaction
(D) The partial pressures of the reactants and the products at 100° C
(E) The entropies of formation for the reactants and the products at 100° C

Ⓐ Ⓑ Ⓒ Ⓓ Ⓔ

124. In the water molecule, the total number of normal vibrational modes is

(A) 1 (B) 2 (C) 3 (D) 4 (E) 5

Ⓐ Ⓑ Ⓒ Ⓓ Ⓔ

125. Which of the following reactions as written is most likely to have a large positive $\triangle S$ of reaction?

(A) $Zn(s) + Cu^{2+}(aq) \rightleftarrows Cu(s) + Zn^{2+}(aq)$

(B) $Zn(s) + 2\,H^{+}(aq) \rightleftarrows H_2(g) + Zn^{2+}(aq)$

(C) $H_2(g) + Cu^{2+}(aq) \rightleftarrows Cu(s) + 2\,H^{+}(aq)$

(D) $8\,Zn(s) + 5\,Cu(s) \rightleftarrows Cu_5Zn_8(s)$

(E) $Zn(s) + H_2(g) \rightleftarrows ZnH_2(s)$

Ⓐ Ⓑ Ⓒ Ⓓ Ⓔ

126. When the natural logarithm of the rate constant k of a reaction is plotted against the reciprocal of the absolute temperature, the slope of the line is related directly to the

(A) reaction energy
(B) free energy
(C) isothermally available energy
(D) activation energy
(E) energy of vaporization of the products

Ⓐ Ⓑ Ⓒ Ⓓ Ⓔ

127.

$[X]_0$	$[Y]_0$	Time Required for [Z] to Increase by 0.0050 mole per liter
0.10M	0.10M	72 sec
0.20M	0.10M	18 sec
0.20M	0.05M	36 sec

The rate data for the net reaction $X + 2Y \rightarrow 3Z$ were obtained at 25° C. The initial rate of increase in [Z] is

(A) first order in both X and Y
(B) first order in X and zero order in Y
(C) second order in X and first order in Y
(D) fourth order in X and second order in Y
(E) second order overall

Ⓐ Ⓑ Ⓒ Ⓓ Ⓔ

128. It is believed that the chemical bond in the hydrogen molecule ion, H_2^+, forms because the

(A) nuclear spins are paired
(B) system has the same number of protons and electrons as He^+
(C) electron is delocalized with a resulting decrease in kinetic energy
(D) potential energy of the system is less than that of the separate parts
(E) reduced mass of the system is identical with that of Li^+

Ⓐ Ⓑ Ⓒ Ⓓ Ⓔ

129.

Salt	KCl	KNO_3	HCl	NaOAc	NaCl
$\Lambda^\circ \dfrac{cm^2}{\text{ohm-equiv.}}$	149.9	145.0	426.2	91.0	126.5

On the basis of the equivalent conductances of the electrolytes listed above at infinite dilution in H_2O at 25° C, one can be certain that which of the following statements is correct?

(A) $\Lambda^\circ_{KCl} - \Lambda^\circ_{HCl} = \Lambda^\circ_{KNO_3} - \Lambda^\circ_{NaCl}$

(B) $\Lambda^\circ_{HNO_3} = 390.7$

(C) $\Lambda^\circ_{HOAc} = \Lambda^\circ_{NaOAc} + \Lambda^\circ_{HCl} - \Lambda^\circ_{NaCl}$

(D) $\lambda^\circ_{H^+} = 325.2$, the limiting equivalent conductance of H^+

(E) $\Lambda^\circ_{KNO_3} - \Lambda^\circ_{HCl} = \Lambda^\circ_{NaOAc} - \Lambda^\circ_{KCl}$

Ⓐ Ⓑ Ⓒ Ⓓ Ⓔ

130. Which of the following graphs best represents the Maxwell distribution law for molecular speeds for a gas sample at a temperature of about 300 K ? (f_S is the fraction of molecules with speed between S and $S + dS$.)

(A)

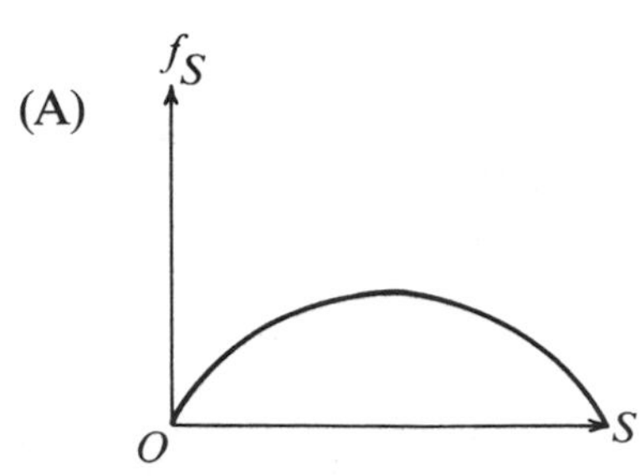

(B)

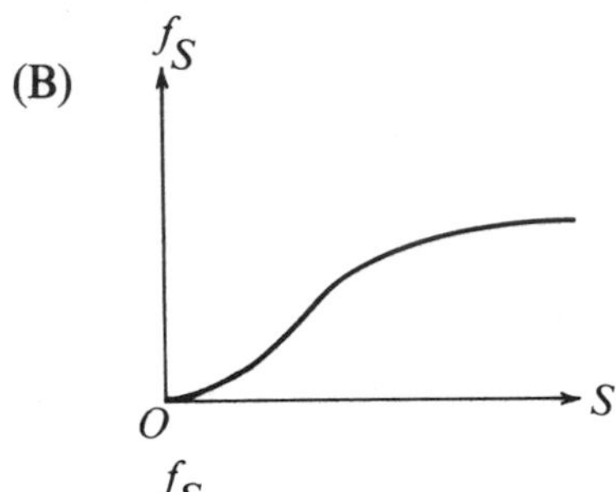

(C)

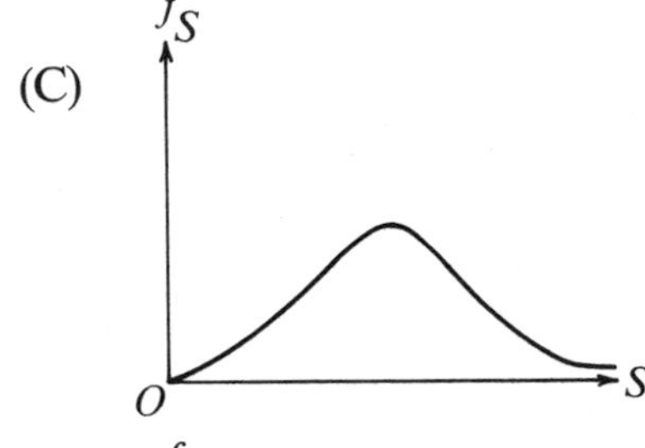

(D)

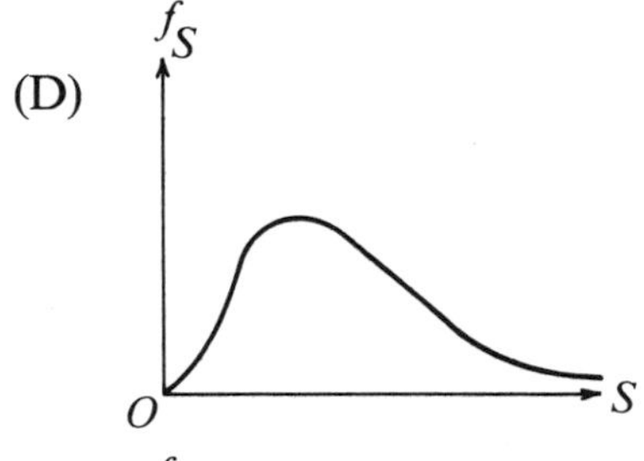

(E)

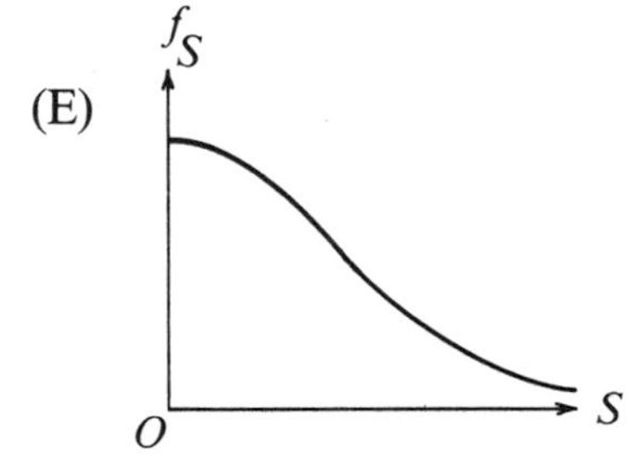

Ⓐ Ⓑ Ⓒ Ⓓ Ⓔ

131. In comparison with HCl, the frequency for the transition of DCl from the ground vibrational state to the first excited vibrational state is

(A) higher for DCl
(B) lower for DCl
(C) sometimes higher and sometimes lower for DCl, depending on the temperature and the concentration
(D) the same for both DCl and HCl
(E) not determinable

Ⓐ Ⓑ Ⓒ Ⓓ Ⓔ

132. An electron in an isolated atom may be described in terms of the four quantum numbers, n, ℓ, m_ℓ, and m_s. The value of m_ℓ for the electron most easily removed from a gaseous atom of an alkaline earth metal is

(A) the same as the maximum value of n for that element
(B) any number (including zero) from $-(n - 1)$ to $+(n - 1)$
(C) any positive number from 1 to $(n - 1)$
(D) $+\frac{1}{2}$
(E) 0

Ⓐ Ⓑ Ⓒ Ⓓ Ⓔ

133.

$$2A \xrightarrow{k} \text{products}$$

The reaction shown above is second order with respect to the concentration of A, denoted [A]. To obtain a straight line with a slope equal to the rate constant, k, one should plot what as a function of time?

(A) 2[A] (B) $[A]^2$ (C) 1/[A] (D) $1/[A]^2$ (E) ln[A]

Ⓐ Ⓑ Ⓒ Ⓓ Ⓔ

134. The shape of a molecule of very high molecular weight in solution can be determined best by which of the following experimental techniques?

(A) Microwave spectroscopy
(B) Osmometry
(C) Nuclear magnetic resonance spectroscopy
(D) Light scattering
(E) Electrophoresis

Ⓐ Ⓑ Ⓒ Ⓓ Ⓔ

135. For a mole of molecules, the thermal energy predicted by quantum theory agrees best with the energy predicted by classical theory for those cases in which the

(A) allowed energy levels are closely spaced compared to kT
(B) allowed energy levels are widely spaced compared to kT
(C) system is at very low temperature
(D) total energy is contributed mostly by vibrational motion
(E) model used is a particle confined to a box approximately 1 cubic angstrom or less in volume

Ⓐ Ⓑ Ⓒ Ⓓ Ⓔ

136. $$\ln P = -\frac{\Delta H}{RT} + \text{constant}$$

The equation above gives the

(A) relationship between the pressure and the temperature for a real gas
(B) relationship between the vapor pressure and the temperature for a condensed phase-vapor phase equilibrium
(C) pressure temperature dependence of solid-solid, liquid-liquid, and solid-liquid equilibria
(D) dependence between the temperature, pressure, and concentration of a dissolved gas in a liquid
(E) enthalpy change accompanying the melting of a solid as a function of pressure and temperature

Ⓐ Ⓑ Ⓒ Ⓓ Ⓔ

137. The interpretation of a unimolecular gas-phase dissociation reaction involves all of the following assumptions EXCEPT:

(A) Molecules are activated by bimolecular collisions.
(B) Only activated molecules can dissociate.
(C) The concentration of activated molecules is constant.
(D) Activated molecules are deactivated by bimolecular collisions.
(E) The reaction rate is larger than the rate of deactivation by collisions.

Ⓐ Ⓑ Ⓒ Ⓓ Ⓔ

138. $Zn(s) + 2\ HCl\ (1\ M) = ZnCl_2\ (1\ M) + H_2\ (1\ atm.)$

$$\Delta H° = -153 \text{ kilojoules/mole}$$

$$\Delta S° = -17.3 \text{ joules/K} \cdot \text{mole}$$

The chemical reaction above can be carried out reversibly in a voltaic cell. If the standard enthalpy and entropy changes for the reaction at 298 K are those given above, the maximum electrical work available from the reaction at this temperature, per mole of H_2 produced, is most nearly

(A) 158,000 J (B) 153,000 J (C) 148,000 J (D) 5,150 J (E) 2,600 J

Ⓐ Ⓑ Ⓒ Ⓓ Ⓔ

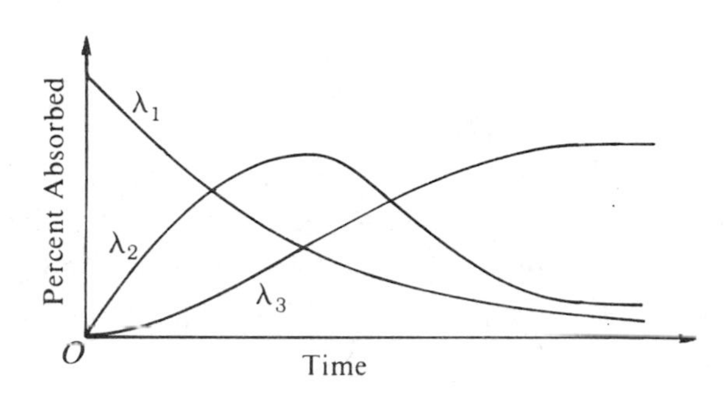

139. The rate of a reaction between A + B was followed spectroscopically at three different wavelengths, and the percent absorption at each wavelength was plotted as a function of time. Which of the following mechanisms is consistent with the spectral data?

(A) A + B → C
(B) A + B → D, D → F
(C) A + B → F, A + B → G
(D) A + B → H, H + A → I
(E) None of the above

Ⓐ Ⓑ Ⓒ Ⓓ Ⓔ

140. What are the conditions favoring a large equilibrium constant for the gas phase isomerization A = B ?

	Partition Function	Ground-State Energy
(A)	Larger for B than for A	Lower for B than for A
(B)	Smaller for B than for A	Lower for B than for A
(C)	Larger for B than for A	Higher for B than for A
(D)	Smaller for B than for A	Higher for B than for A
(E)	Larger for B than for A	Equal for A and B

Ⓐ Ⓑ Ⓒ Ⓓ Ⓔ

141. Let $\triangle G°$ be the change in the standard Gibbs free energy accompanying an oxidation-reduction reaction and K and $E°$ be the corresponding equilibrium constant and standard potential, respectively. Which of the following is a consistent set of relations?

(A) $\triangle G° > 0$ $\quad E° < 0$ $\quad K < 1$

(B) $\triangle G° > 0$ $\quad E° > 0$ $\quad K > 1$

(C) $\triangle G° > 0$ $\quad E° < 0$ $\quad K > 1$

(D) $\triangle G° < 0$ $\quad E° < 0$ $\quad K > 1$

(E) $\triangle G° < 0$ $\quad E° > 0$ $\quad K < 1$

Ⓐ Ⓑ Ⓒ Ⓓ Ⓔ

142.

$$\left[\frac{1}{N_0}\frac{dN}{dc} = \rho_{(c)}\right]$$

Maxwell's distribution law for molecular speeds may be given by the equation above where N_0 = total number of molecules in the sample, $c \equiv$ molecular speed,

$$\rho_{(c)} = 4\pi\left(\frac{m}{2\pi kt}\right)^{\frac{3}{2}} \exp\left(-mc^2/2kT\right)c^2,$$

and $m \equiv$ molecular mass. To compute the most probable speed of a molecule in the gas sample, one would

(A) solve for the c which satisfies the expression $\frac{\partial \rho(c)}{\partial c} = 0$

(B) evaluate the integral $\int_0^{\infty} \rho_{(c)} dc$

(C) evaluate the integral $\int_0^{\infty} \rho_{(c)} c dc$

(D) evaluate the integral $\int_{-\infty}^{+\infty} \rho_{(c)} dc$

(E) evaluate the integral $\int_{-\infty}^{\infty} \rho_{(c)} c dc$

Ⓐ Ⓑ Ⓒ Ⓓ Ⓔ

143.

	ΔH_f° (kilocalories per mole)	S° (calories per mole · K)
graphite	0.0	1.37
diamond	0.45	0.57

The values for the standard enthalpy of formation and the standard entropy of graphite and diamond at $T = 298$ K are given in the table above. At this temperature and a pressure of 1 atmosphere, which of the following is true?

(A) Graphite is more stable than diamond from the standpoint of both enthalpy and entropy.
(B) Diamond is more stable than graphite from the standpoint of both entropy and enthalpy.
(C) Graphite is more stable than diamond from the standpoint of entropy, but diamond is more stable than graphite from the standpoint of enthalpy.
(D) Graphite is more stable than diamond from the standpoint of enthalpy, but diamond is more stable than graphite from the standpoint of entropy.
(E) The standard molar Gibbs free energy of graphite is greater than that of diamond.

144. The freezing point of water is lowered by the addition of a soluble substance such as sodium chloride. This lowering is considered to be a consequence of the fact that

(A) the partial molal volume of ice is greater than the partial molal volume of liquid water at the freezing point of the solution
(B) the vapor pressure of pure ice is less than that of the water in solution at the normal freezing point of pure ice
(C) the chemical potential of water in the solution at the normal freezing point of water is less than that of pure ice
(D) sodium chloride dissociates into ions when it dissolves in water
(E) the dissolving of sodium chloride in water is an exothermic process

145. A solution containing 0.10 mole of B and 0.90 mole of A exhibits a positive deviation from Raoult's law. The activity coefficient of solvent A is

(A) negative
(B) zero
(C) between zero and 1.0
(D) 1.0
(E) greater than 1.0

Ⓐ Ⓑ Ⓒ Ⓓ Ⓔ

SAMPLE QUESTIONS ANSWER KEY

Analytical Chemistry	Inorganic Chemistry	Organic Chemistry	Physical Chemistry
1. D	24. A	58. E	102. C
2. C	25. C	59. A	103. A
3. A	26. D	60. D	104. D
4. D	27. B	61. E	105. C
5. B	28. D	62. B	106. C
6. E	29. C	63. E	107. B
7. C	30. A	64. B	108. B
8. C	31. C	65. C	109. B
9. E	32. E	66. A	110. B
10. C	33. A	67. B	111. C
11. A	34. B	68. E	112. E
12. C	35. C	69. D	113. D
13. B	36. B	70. C	114. A
14. B	37. C	71. C	115. B
15. D	38. D	72. E	116. D
16. B	39. B	73. C	117. B
17. E	40. E	74. A	118. A
18. C	41. D	75. A	119. A
19. E	42. B	76. E	120. C
20. E	43. A	77. C	121. D
21. A	44. C	78. E	122. E
22. D	45. A	79. D	123. B
23. E	46. D	80. B	124. C
	47. B	81. D	125. B
	48. E	82. E	126. D
	49. A	83. B	127. C
	50. D	84. D	128. D
	51. E	85. E	129. C
	52. E	86. C	130. D
	53. E	87. D	131. B
	54. E	88. D	132. E
	55. C	89. B	133. C
	56. D	90. D	134. D
	57. D	91. A	135. A
		92. E	136. B
		93. B	137. E
		94. C	138. C
		95. B	139. B
		96. E	140. A
		97. A	141. A
		98. D	142. A
		99. B	143. A
		100. D	144. C
		101. E	145. E

TAKING THE TEST

TEST-TAKING STRATEGY

Presumably, if you are about to take the GRE Chemistry Test, you are nearing completion of or have completed an undergraduate curriculum in that subject. A general review of your curriculum is probably the best preparation for taking the test. Because the level of difficulty of the test is set to provide reliable measurement over a broad range of subject matter, you are not expected to be able to answer every question correctly.

You are strongly urged to work through some of the sample questions preceding this section. After you have evaluated your performance within the content categories, you may determine that a review of certain courses would be to your benefit.

In preparing to take the full-length Chemistry Test, it is important that you become thoroughly familiar with the directions provided in the full-length test included in this book. For this test, your score will be determined by subtracting one-fourth the number of incorrect answers from the number of correct answers. Questions for which you mark no answer or more than one answer are not counted in scoring. If you have some knowledge of a question and are able to rule out one or more of the answer choices as incorrect, your chances of selecting the correct answer are improved, and answering such questions is likely to improve your score. It is unlikely that pure guessing will raise your score; it may lower your score.

Work as rapidly as you can without being careless. *This includes checking frequently to make sure you are marking your answers in the appropriate rows on your answer sheet.* Since no question carries greater weight than any other, do not waste time pondering individual questions you find extremely difficult or unfamiliar.

You may find it advantageous to go through the test a first time quite rapidly, stopping only to answer those questions of which you are confident. Then go back and answer the questions that require greater thought, concluding with the very difficult questions, if you have time.

HOW TO SCORE YOUR TEST

Total Subject Test scores are reported as three-digit scaled scores with the third digit always zero. The maximum possible range for all Subject Test total scores is from 200 to 990. The actual range of scores for a particular Subject Test, however, may be smaller. Chemistry Test scores typically range from 440 to 920. The range for different editions of a given test may vary because different editions are not of precisely the same difficulty. The differences in ranges among different editions of a given test, however, usually are small. This should be taken into account, especially when comparing two very high scores. In general, differences between scores at the 99th percentile should be ignored. **The score conversions table provided shows the score range for this edition of the test only.**

The work sheet on page 66 lists the correct answers to the questions. Columns are provided for you to mark whether you chose the correct (C) answer or an incorrect (I) answer to each question. Draw a line across any question you omitted, because it is not counted in the scoring. At the bottom of the page, enter the total number correct and the total number incorrect. Divide the total incorrect by 4 and subtract the resulting number from the total correct. This is the adjustment made for guessing. Then round the result to the nearest whole number. This will give you your raw total score. Use the total score conversion table to find the scaled total score that corresponds to your raw total score.

Example: Suppose you chose the correct answers to 75 questions and incorrect answers to 46. Dividing 46 by 4 yields 11.5. Subtracting 11.5 from 75 equals 63.5, which is rounded to 64. The raw score of 64 corresponds to a scaled score of 660.

WORK SHEET for the Chemistry Test, Form GR9027 ONLY
Answer Key and Percentages* of Examinees Answering Each Question Correctly

QUESTION Number	QUESTION Answer	P+	TOTAL C	TOTAL I
1	D	62		
2	C	85		
3	B	74		
4	D	61		
5	C	68		
6	E	81		
7	A	57		
8	A	33		
9	D	30		
10	B	55		
11	E	69		
12	A	77		
13	E	90		
14	D	59		
15	C	65		
16	D	71		
17	B	39		
18	D	60		
19	A	58		
20	B	78		
21	A	60		
22	E	65		
23	C	67		
24	B	73		
25	E	75		
26	E	68		
27	D	44		
28	B	29		
29	D	73		
30	B	52		
31	D	65		
32	C	52		
33	A	55		
34	A	52		
35	C	46		
36	C	61		
37	D	56		
38	C	55		
39	B	50		
40	B	55		
41	C	55		
42	B	42		
43	B	60		
44	A	55		
45	D	43		
46	B	51		
47	A	64		
48	E	77		
49	B	70		
50	E	51		
51	C	60		
52	E	49		
53	C	26		
54	B	55		
55	D	18		

QUESTION Number	QUESTION Answer	P+	TOTAL C	TOTAL I
56	A	39		
57	B	9		
58	E	36		
59	C	33		
60	C	34		
61	C	80		
62	C	53		
63	D	24		
64	C	80		
65	D	65		
66	B	45		
67	C	44		
68	E	43		
69	B	45		
70	A	42		
71	A	51		
72	E	43		
73	A	27		
74	E	46		
75	B	26		
76	A	59		
77	B	42		
78	E	27		
79	C	40		
80	E	34		
81	E	39		
82	D	44		
83	B	30		
84	D	48		
85	B	63		
86	D	36		
87	B	65		
88	D	36		
89	D	40		
90	B	20		
91	C	52		
92	A	38		
93	A	45		
94	B	38		
95	D	30		
96	C	41		
97	A	39		
98	E	80		
99	B	41		
100	C	46		
101	E	28		
102	C	40		
103	A	41		
104	A	31		
105	D	22		
106	C	57		
107	B	61		
108	D	30		
109	E	43		
110	A	30		

QUESTION Number	QUESTION Answer	P+	TOTAL C	TOTAL I
111	D	27		
112	A	27		
113	E	28		
114	A	60		
115	D	21		
116	C	33		
117	E	36		
118	C	31		
119	B	18		
120	C	30		
121	B	39		
122	B	36		
123	C	15		
124	D	25		
125	C	28		
126	A	26		
127	A	50		
128	B	29		
129	E	29		
130	D	37		
131	A	40		
132	D	53		
133	D	19		
134	B	67		
135	E	18		
136	E	45		
137	E	28		
138	C	19		
139	A	48		
140	A	17		
141	C	64		
142	E	23		
143	A	28		
144	B	39		
145	D	50		
146	D	25		
147	B	23		
148	B	54		
149	C	21		
150	D	22		

Correct (C) ______________________

Incorrect (I) ______________________

Total Score:

C – I/4 = ____________

Scaled Score (SS) = ____________

*The P+ column lists the percent of an analysis sample of Chemistry Test examinees who answered each question correctly; this sample consists of October 1989 examinees selected to represent all Chemistry Test examinees tested between October 1, 1986, and September 30, 1989.

SCORE CONVERSIONS AND PERCENTS BELOW*
FOR GRE CHEMISTRY TEST, Form GR9027 ONLY

TOTAL SCORE					
Raw Score	Scaled Score	%	Raw Score	Scaled Score	%
150	980	99	71-73	690	70
147-149	970	99	69-70	680	67
144-146	960	99	66-68	670	64
141-143	950	99	63-65	660	61
139-140	940	99	61-62	650	58
136-138	930	99	58-60	640	55
133-135	920	99	55-57	630	51
131-132	910	99	53-54	620	47
128-130	900	98	50-52	610	44
125-127	890	98	47-49	600	41
123-124	880	97	44-46	590	37
120-122	870	97	42-43	580	33
117-119	860	96	39-41	570	29
115-116	850	96	36-38	560	26
112-114	840	95	34-35	550	22
109-111	830	94	31-33	540	19
106-108	820	93	28-30	530	16
104-105	810	92	26-27	520	14
101-103	800	91	23-25	510	11
98-100	790	90	20-22	500	9
96-97	780	89	18-19	490	7
93-95	770	87	15-17	480	5
90-92	760	85	12-14	470	4
88-89	750	84	9-11	460	3
85-87	740	82	7-8	450	2
82-84	730	79	4-6	440	1
80-81	720	77	1-3	430	1
77-79	710	75	0	420	0
74-76	700	73			

*Percent scoring below the scaled score based on the performance of 14,344 examinees who took the GRE Subject Test in Chemistry between October 1, 1986, and September 30, 1989.

EVALUATING YOUR PERFORMANCE

Now that you have scored your test, you may wish to compare your performance with the performance of others who took this test. Two kinds of information are provided, both using performance data from GRE Chemistry examinees tested between October 1986 and September 1989. Interpretive data based on the scores earned by examinees tested in this three-year period are to be used by admissions officers in 1990-91.

The first kind of information is based on the performance of a sample of the examinees who took the test in October 1989. This sample was selected to represent the total population of GRE Chemistry examinees tested between October 1986 and September 1989. On the work sheet you used to determine your score is a column labeled "P+." The numbers in this column indicate the percent of the examinees in this sample who answered each question correctly. You may use these numbers as a guide for evaluating your performance on each test question.

Also included, for each scaled score, is the percent of examinees tested between October 1986 and September 1989 who received lower scores. These percents appear in the score conversions table in a column to the right of the scaled scores. For example, in the percent column opposite the scaled score of 660 is the percent 61. This means that 61 percent of the Chemistry Test examinees tested between October 1986 and September 1989 scored lower than 660. To compare yourself with this population, look at the percent next to the scaled score you earned on the practice test. This number is a reasonable indication of your rank among GRE Chemistry Test examinees if you followed the test-taking suggestions in this practice book.

It is important to realize that the conditions under which you tested yourself were not exactly the same as those you will encounter at a test center. It is impossible to predict how different test-taking conditions will affect test performance, and this is only one factor that may account for differences between your practice test scores and your actual test scores. By comparing your performance on this practice test with the performance of other GRE Chemistry Test examinees, however, you will be able to determine your strengths and weaknesses and can then plan a program of study to prepare yourself for taking the Chemistry Test under standard conditions.

Before you start timing yourself on the test that follows, we suggest that you remove an answer sheet (pages 111 to 118) and turn first to the back cover of the test book (page 110), as you will do at the test center, and follow the instructions for completing the identification areas of the answer sheet. Then read the inside back cover instructions (page 109). When you are ready to begin the test, note the time and start marking your answers to the questions on the answer sheet.

FORM GR9027

27

THE GRADUATE RECORD EXAMINATIONS

CHEMISTRY TEST

Do not break the seal until you are told to do so.

The contents of this test are confidential. Disclosure or reproduction of any portion of it is prohibited.

THIS TEST BOOK MUST NOT BE TAKEN FROM THE ROOM.

Material in the following table may be useful in answering the questions in this examination.

PERIODIC CHART OF THE ELEMENTS

1 **H** 1.0079																	2 **He** 4.003
3 **Li** 6.941	4 **Be** 9.012											5 **B** 10.81	6 **C** 12.011	7 **N** 14.007	8 **O** 16.00	9 **F** 19.00	10 **Ne** 20.179
11 **Na** 22.99	12 **Mg** 24.30											13 **Al** 26.98	14 **Si** 28.09	15 **P** 30.974	16 **S** 32.06	17 **Cl** 35.453	18 **Ar** 39.948
19 **K** 39.10	20 **Ca** 40.08	21 **Sc** 44.96	22 **Ti** 47.90	23 **V** 50.94	24 **Cr** 52.00	25 **Mn** 54.94	26 **Fe** 55.85	27 **Co** 58.93	28 **Ni** 58.70	29 **Cu** 63.55	30 **Zn** 65.38	31 **Ga** 69.72	32 **Ge** 72.59	33 **As** 74.92	34 **Se** 78.96	35 **Br** 79.90	36 **Kr** 83.80
37 **Rb** 85.47	38 **Sr** 87.62	39 **Y** 88.91	40 **Zr** 91.22	41 **Nb** 92.91	42 **Mo** 95.94	43 **Tc** (97)	44 **Ru** 101.1	45 **Rh** 102.91	46 **Pd** 106.4	47 **Ag** 107.868	48 **Cd** 112.41	49 **In** 114.82	50 **Sn** 118.7	51 **Sb** 121.75	52 **Te** 127.60	53 **I** 126.90	54 **Xe** 131.30
55 **Cs** 132.91	56 **Ba** 137.33	57 ***La** 138.91	72 **Hf** 178.49	73 **Ta** 180.95	74 **W** 183.85	75 **Re** 186.21	76 **Os** 190.2	77 **Ir** 192.2	78 **Pt** 195.09	79 **Au** 196.97	80 **Hg** 200.59	81 **Tl** 204.37	82 **Pb** 207.2	83 **Bi** 208.98	84 **Po** (209)	85 **At** (210)	86 **Rn** (222)
87 **Fr** (223)	88 **Ra** (226)	89 †**Ac** (227)															

*Lanthanum Series

58 **Ce** 140.12	59 **Pr** 140.91	60 **Nd** 144.24	61 **Pm** (145)	62 **Sm** 150.4	63 **Eu** 152.0	64 **Gd** 157.25	65 **Tb** 158.93	66 **Dy** 162.50	67 **Ho** 164.93	68 **Er** 167.26	69 **Tm** 168.93	70 **Yb** 173.04	71 **Lu** 174.97

†Actinium Series

90 **Th** 232.0	91 **Pa** 231.0	92 **U** 238.03	93 **Np** 237.0	94 **Pu** (244)	95 **Am** (243)	96 **Cm** (247)	97 **Bk** (247)	98 **Cf** (251)	99 **Es** (252)	100 **Fm** (257)	101 **Md** (258)	102 **No** (259)	103 **Lr** (260)

GO ON TO THE NEXT PAGE.

CHEMISTRY TEST

Time—170 minutes

150 Questions

Directions: Each of the questions or incomplete statements below is followed by five suggested answers or completions. Select the one that is best in each case and then fill in the corresponding space on the answer sheet.

Note: Solutions are aqueous unless otherwise specified.

1. Hydrogen bonding is exhibited by which of the following substances?

(A) KH
(B) H_2
(C) CH_4
(D) HF
(E) AsH_3

2. The number of neutrons in an atom of $^{207}_{82}Pb$ is

(A) 41 (B) 82 (C) 125
(D) 207 (E) 289

3. In the NO_2^- ion, the nitrogen-oxygen bond length corresponds to a

(A) single bond
(B) bond intermediate between a single and double bond
(C) double bond
(D) bond intermediate between a double and triple bond
(E) triple bond

4. Addition of solid sodium acetate to aqueous acetic acid results in

(A) an increase in the ionization constant of the acetic acid
(B) no change in the OH^- ion concentration of the solution
(C) a decrease in the OH^- ion concentration of the solution
(D) a decrease in the H^+ ion concentration of the solution
(E) an increase in the degree of ionization of the acetic acid

5. The highest fluoride formed by carbon (atomic number 6) is CF_4, whereas the highest fluoride formed by silicon (atomic number 14) is SiF_6^{2-}. This is primarily because

(A) silicon has a lower ionization energy than carbon has
(B) silicon is more metallic than carbon is
(C) silicon atoms possess empty low-lying *d* orbitals, whereas carbon does not
(D) the *d* orbitals of uncombined carbon atoms are filled, whereas those of uncombined silicon atoms are empty
(E) the *d* orbitals of uncombined silicon atoms are filled, whereas those of uncombined carbon atoms are empty

6.

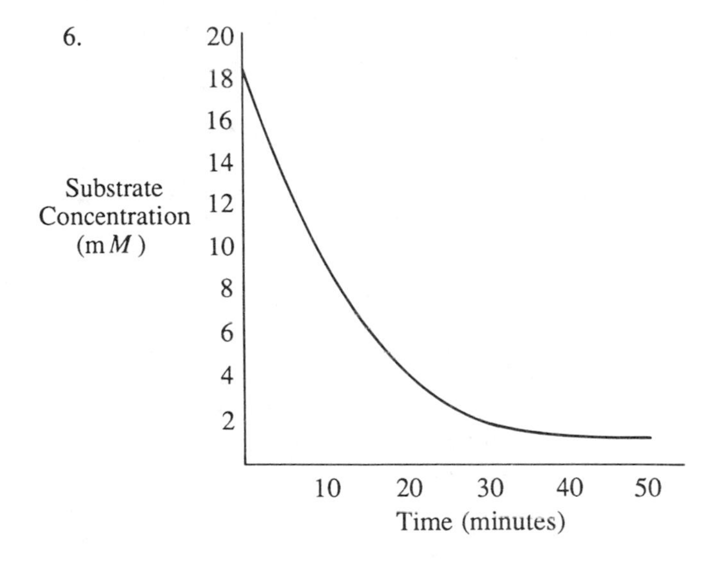

The graph above illustrates the course of most enzymatic reactions as they occur in the laboratory. Enzyme is mixed with the substrate at time zero. The concentration of substrate declines rapidly at first, but is never reduced to zero probably because

(A) the enzyme is unstable and disappears within an hour
(B) the end products occupy the active site of the enzyme
(C) the heat of the reaction brings it to a halt
(D) cofactors for the reaction are used up
(E) an equilibrium is reached

GO ON TO THE NEXT PAGE.

7. Transmittance in spectroscopic analysis is a function primarily of the

(A) optical path
(B) velocity of the light
(C) reference solvent
(D) monochromator
(E) barometric pressure

8. Which of the following techniques is LEAST suited for quantitative measurement of trace organic compounds in the environment?

(A) Nuclear magnetic resonance spectroscopy
(B) Fluorescence spectroscopy
(C) Mass spectroscopy
(D) Gas chromatography
(E) Liquid chromatography

9. The pH at the equivalence point in the titration of a 0.010-molar solution of a weak acid, HA, ($pK_a = 4.0$) with a 0.010-molar solution of NaOH is

(A) 4.0
(B) greater than 4.0 and less than 7.0
(C) 7.0
(D) greater than 7.0
(E) not calculable because the initial volume of the weak acid solution is not given

10. Excess silver nitrate is added to a known volume of a solution containing only sodium chloride and potassium bromide. To determine the composition of the original solution, it is necessary and sufficient to know the

(A) weight of the original solution and the weight of the precipitate
(B) total weight of the solute in the original solution and the weight of the precipitate
(C) molarity of the silver nitrate solution used
(D) density of the original solution at a specified temperature
(E) density and the specific conductance of the original solution

11. Peptide bonding results in the formation of an

(A) ester (B) aldehyde (C) ether
(D) acetal (E) amide

12. The reaction of nitrobenzene with an electrophilic reagent proceeds less rapidly than the reaction of benzene with the same reagent because the

(A) nitro group reduces the electron density of the aromatic ring to which it is attached
(B) electrophilic reagent complexes with the nitro group and thus becomes less available for reaction at a site on the ring
(C) bulk of the nitro group shields the aromatic ring from attack by the reagent
(D) charges on the oxygen atoms of the nitro group repel the electrophilic reagent
(E) nitro group increases the electron density of the ring to such an extent that the ring loses its aromatic properties

13. What experimental parameter can increase the rate of a chemical reaction by lowering the energy of activation?

(A) Increasing the concentration of reactants
(B) Elevating the reaction temperature
(C) Removing the products from the reaction mixture by distillation
(D) Improving the mixing by continuous agitation of the reaction mixture
(E) Introducing a catalyst into the reaction

GO ON TO THE NEXT PAGE.

14. The chlorination of neopentane is initiated by light and yields neopentyl chloride. The most likely mechanism for this reaction is

(A) $Cl_2 \xrightarrow{\text{light}} Cl^+ + Cl^-$

$Cl^+ + (CH_3)_3CCH_3 \longrightarrow (CH_3)_3CCH_2Cl + H^+$

$H^+ + Cl^- \longrightarrow HCl$

(B) $Cl_2 \xrightarrow{\text{light}} Cl^+ + Cl^-$

$Cl^+ + (CH_3)_3CCH_3 \longrightarrow (CH_3)_3C{-}\overset{+}{C}H_2 + HCl$

$(CH_3)_3C{-}\overset{+}{C}H_2 + Cl^- \longrightarrow (CH_3)_3CCH_2Cl$

(C) $Cl_2 \xrightarrow{\text{light}} 2\ Cl\cdot$

$Cl\cdot + (CH_3)_3CCH_3 \longrightarrow (CH_3)_3CCH_2Cl + H\cdot$

$H\cdot + Cl\cdot \longrightarrow HCl$

(D) $Cl_2 \xrightarrow{\text{light}} 2\ Cl\cdot$

$Cl\cdot + (CH_3)_3CCH_3 \longrightarrow (CH_3)_3C{-}\dot{C}H_2 + HCl$

$(CH_3)_3C{-}\dot{C}H_2 + Cl_2 \longrightarrow (CH_3)_3CCH_2Cl + Cl\cdot$

(E) $Cl_2 \xrightarrow{\text{light}} 2\ Cl\cdot$

$Cl\cdot + (CH_3)_3CCH_3 \longrightarrow (CH_3)_3C{-}\dot{C}H_2 + HCl$

$(CH_3)_3C{-}\dot{C}H_2 + Cl\cdot \longrightarrow (CH_3)_3CCH_2Cl$

15.

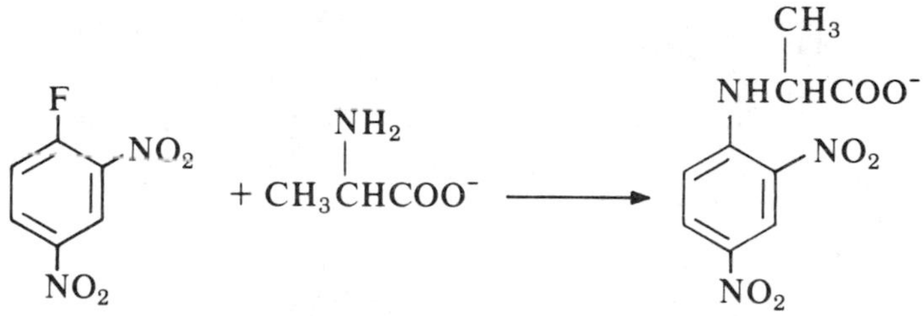

Which of the following terms or phrases best describes the type of reaction illustrated above?

(A) Electrophilic addition
(B) Electrophilic aromatic substitution
(C) Nucleophilic aromatic substitution
(D) Nucleophilic addition
(E) β-Elimination

GO ON TO THE NEXT PAGE.

16. If symbols are used as defined, which of the following statements gives a correct association?

U = crystal lattice energy

I = ionization energy

S = enthalpy of sublimation

D = enthalpy of dissociation

(A) For the process $Cl^- \rightarrow Cl + e^-$, the enthalpy of reaction is I.
(B) For the process $Na^+ + e^- \rightarrow Na$, the enthalpy of reaction is U.
(C) For the process $Na + \frac{1}{2}Cl_2 \rightarrow NaCl$, the enthalpy of reaction is S.
(D) For the process $Cl_2(gas) \rightarrow 2\ Cl(gas)$, the enthalpy of reaction is D.
(E) For the process $Na(solid) \rightarrow Na(gas)$, the energy required for the reaction is D.

17. The coordination number for the cations in a crystal of CaF_2 is 8. The coordination number for the anions in this crystal is

(A) 2
(B) 4
(C) 6
(D) 8
(E) 16

18.

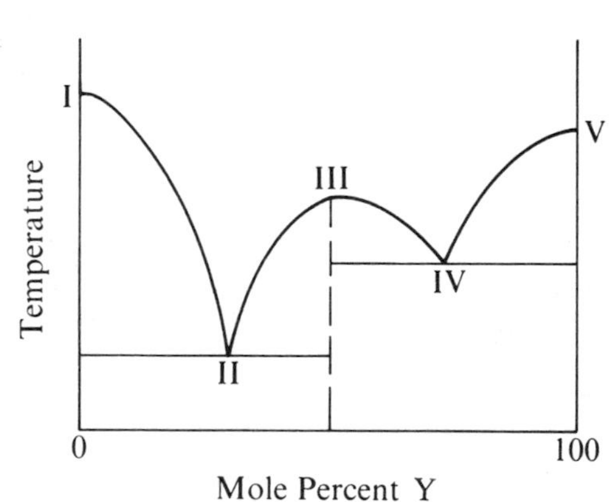

A phase diagram for a two-component system of substances X and Y is shown above. A eutectic point is represented by which of the following?

(A) II only
(B) III only
(C) I and V only
(D) II and IV only
(E) II, III, and IV

19. Photosensitizers for a reaction (such as those used on photographic film) are effective because they

(A) absorb light that would not be absorbed by the reagents themselves and thus make the energy available for reaction
(B) disperse excess light energy that otherwise would be absorbed by the reagents and would interfere with the reaction
(C) convert heat energy into light that can cause reaction
(D) convert the chemical energy of the reaction into light, which is then emitted
(E) refract light into the reagents

20.

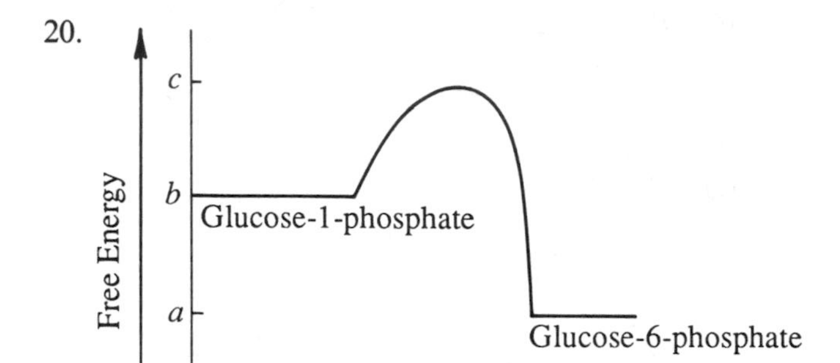

For the reaction

glucose-1-phosphate → glucose-6-phosphate

energy a minus b in the diagram above is the

(A) free energy change for the reaction and is positive
(B) free energy change for the reaction and is negative
(C) activation energy for the reaction and is positive
(D) activation energy for the reaction and is negative
(E) enthalpy change for the reaction and is positive

21. Alcohols of the type RR′R″COH may be prepared by the action of a Grignard reagent on

(A) a ketone
(B) a secondary alcohol
(C) an ester
(D) an aldehyde
(E) a nitrile

GO ON TO THE NEXT PAGE.

22. When each of the following compounds undergoes nitration, which would be expected to give the lowest ratio of moles of *ortho* isomer to moles of *para* isomer?

(A) $C_6H_5-CH_2CH_3$

(B) $C_6H_5-CHCl_2$

(C) $C_6H_5-CH_2Br$

(D) $C_6H_5-CH_3$

(E) $C_6H_5-C(CH_3)_3$

23. The S_N2 reaction is known to occur with

(A) racemization
(B) partial inversion
(C) almost complete inversion
(D) retention of configuration
(E) mutarotation

24. Dimerization of cyclopentadiene forming 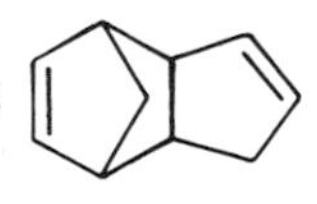 is an example of

(A) a free-radical chain reaction
(B) a Diels-Alder reaction
(C) an aldol condensation
(D) a Friedel-Crafts reaction
(E) a condensation polymerization

---

25. Which of the following compounds would NOT give a single resonance line (a singlet) as the only resonance line in the proton nuclear magnetic resonance spectrum?

(A) $CH_3-CCl_2-CH_3$

(B) $CH_3-CBr(CH_3)-CH_3$

(C) $BrH_2C-CClBr-CH_2Br$

(D) $CH_2=CH_2$

(E) $Br-CH_2-CH_2-I$

GO ON TO THE NEXT PAGE.

26. Which of the following wave numbers corresponds to energy in the infrared region of the electromagnetic spectrum?

(A) 1,700,000 cm^{-1}
(B) 100,000 cm^{-1}
(C) 33,000 cm^{-1}
(D) 20,000 cm^{-1}
(E) 2,000 cm^{-1}

27. Which of the following curves represents the change in pH when 25.0 milliliters of 0.1-molar acetic acid is titrated with 0.1-molar sodium hydroxide?

(A)

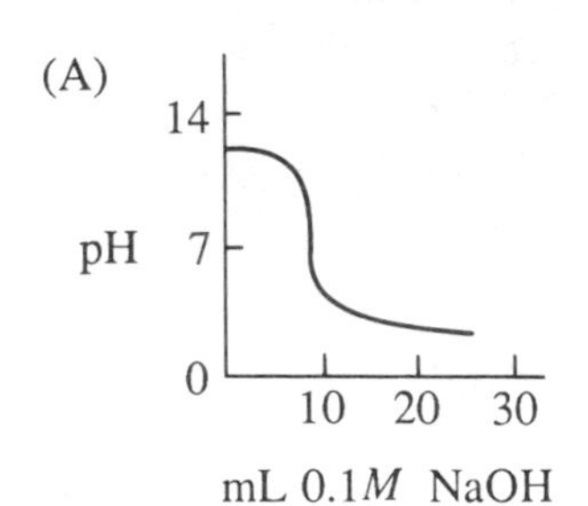

(B)

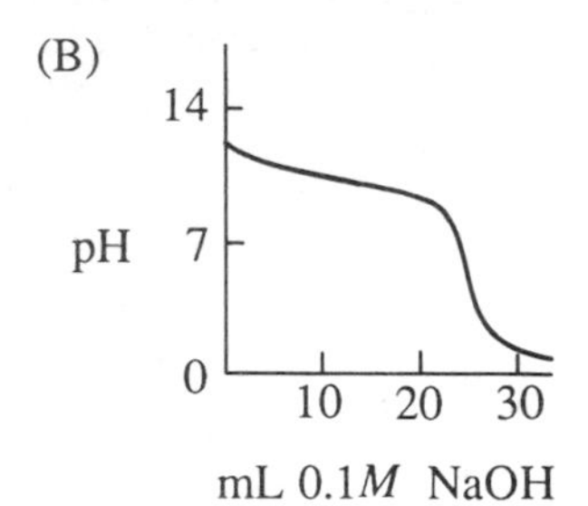

(C)

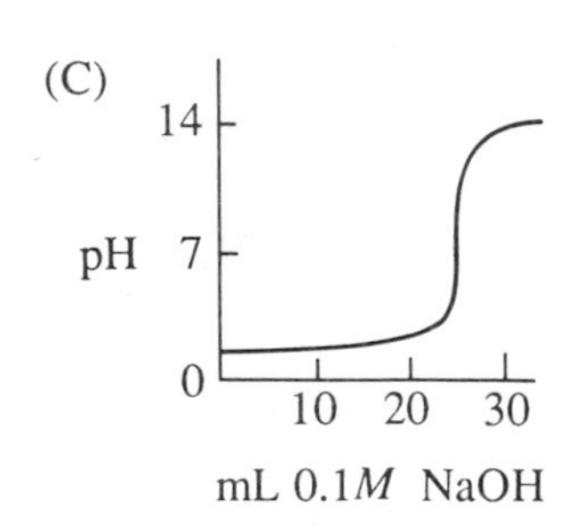

(D)

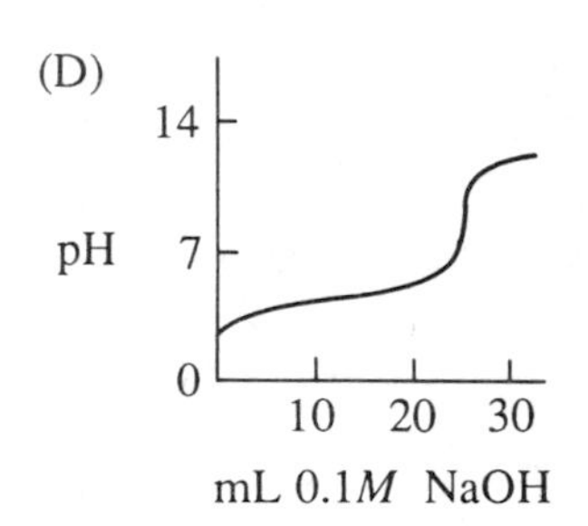

(E)

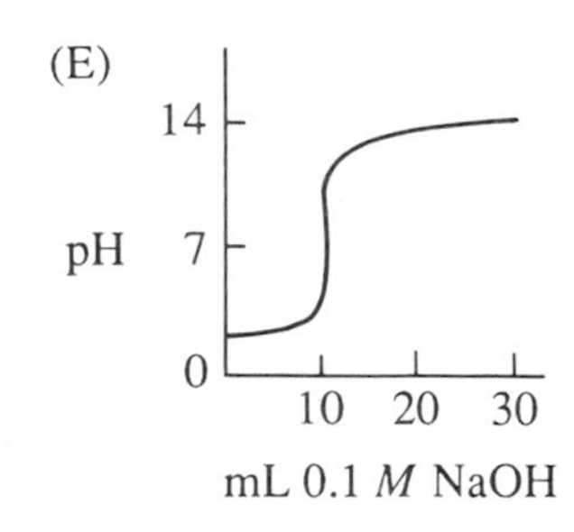

28. Which of the following relationships is correct at the equivalence point in the titration of Fe^{3+} with Sn^{2+} ?

(A) $[Fe^{3+}] = [Sn^{4+}]$

(B) $\frac{[Fe^{3+}]}{[Fe^{2+}]} = \frac{[Sn^{2+}]}{[Sn^{4+}]}$

(C) $[Sn^{2+}] = [Sn^{4+}]$

(D) $\frac{[Sn^{4+}]}{[Sn^{2+}]} = \frac{[Fe^{3+}]}{[Fe^{2+}]}$

(E) $[Fe^{2+}] = [Fe^{3+}]$

29. Absorption of spectral energy in the infrared region is primarily the result of

(A) interaction of the light energy with the nuclei of atoms
(B) excitation of inner electrons in atoms to higher energy levels
(C) excitation of bonding electrons in atoms to higher energy levels
(D) excitation of molecules from one vibrational energy level to a higher one
(E) excitation of atoms from one translational energy level to a higher one

30. All of the following are characteristics of gas chromatography EXCEPT:

(A) A mobile inert gas carries the sample.
(B) The stationary phase is a volatile liquid.
(C) The sample undergoing separation may be a gas, a liquid, or a solid at room temperature.
(D) The separation column may be a capillary without packing material.
(E) The number of plates for separation may be calculated.

31. In the van der Waals equation

$$\left(P + \frac{a}{V^2}\right)(V - b) = RT$$

the term $\frac{a}{V^2}$ is introduced to correct for

(A) the volume occupied by the molecules themselves
(B) the effects of the kinetic energy of the molecules
(C) the momentum changes when molecules collide
(D) the effects of forces of attraction between molecules
(E) statistical variations resulting from the crooked paths traveled by molecules

32. $$\mathcal{E} = \mathcal{E}^\circ - \frac{RT}{nF} \ln Q$$

If the stoichiometric coefficients in the net cell reaction for an electrochemical cell were doubled, which of the following terms in the expression above would also be doubled? (Q = reaction quotient, T = temperature, F = faraday, n = number of faradays needed for the cell reaction, $\mathcal{E}$ = electromotive force)

(A) $\mathcal{E}$
(B) $\mathcal{E}^\circ$
(C) n
(D) Q
(E) F

GO ON TO THE NEXT PAGE.

33. For the spontaneous freezing of supercooled water at a constant temperature of $-20°C$ and 1.0 atmosphere pressure, which of the following is correct? (G = Gibbs energy, S = entropy, H = enthalpy, U = internal energy)

(A) $\Delta G_{system} < 0$

(B) $\Delta S_{system} > 0$

(C) $\Delta H_{system} > 0$

(D) $\Delta U_{surroundings} < 0$

(E) $\Delta S_{surroundings} < 0$

34. When a vibration-rotation spectrum of gaseous HCl is observed under high resolution, each peak is split into two components. This effect results from

(A) the presence of the two isotopes of chlorine
(B) a change in the force constant at the extrema of the vibration
(C) a change in the moment of inertia at the extrema of the vibration
(D) interaction with higher harmonics of the vibration
(E) violation of the selection rules

35. Which of the following is NOT true about the effect of an increase in temperature on the distribution of molecular speeds in a gas?

(A) The average speed shifts to larger values.
(B) The most probable speed increases.
(C) The fraction of the molecules with the most probable speed increases.
(D) The area under the distribution curve remains the same.
(E) The distribution becomes broader.

36. Which of the following molecules or ions does NOT have a linear structure?

(A) NO_2^+ (B) HCN (C) H_2S
(D) CO_2 (E) ICl_2^-

37. Hybridization of d^2sp^3 orbitals is associated with what structure?

(A) Trigonal bipyramidal
(B) Planar
(C) Tetrahedral
(D) Octahedral
(E) Dodecahedral

38. The atomic radius is largest for which of the following elements?

(A) Be (B) Mg (C) Ca (D) Mn (E) Zn

39. Which of the following is a peroxide?

(A) SO_2 (B) BaO_2 (C) MnO_2
(D) SiO_2 (E) SnO_2

40. Bond strengths in the nitrogen, oxygen, and fluorine molecules follow the order $N_2 > O_2 > F_2$. According to the molecular orbital theory, this trend is best explained on the basis of which of the following statements?

(A) The magnitudes of the molar enthalpies of formation of the gaseous atoms decrease in the order $N(g) > O(g) > F(g)$.
(B) The number of electrons in antibonding orbitals increases in the order $N_2 < O_2 < F_2$.
(C) The number of electrons in bonding molecular orbitals decreases in the order $N_2 > O_2 > F_2$.
(D) The electronegativities increase in the order $N < O < F$.
(E) The molecular weights increase in the order $N_2 < O_2 < F_2$.

41. The wavelike character of the electron is demonstrated by which of the following?

(A) Photoelectron spectroscopy
(B) Electron spin resonance
(C) Electron diffraction
(D) Electron spectroscopy for chemical analysis
(E) Electron-impact mass spectroscopy

42. $\alpha(1)\alpha(2)$ $\beta(1)\beta(2)$ $\alpha(1)\beta(2)$ $\beta(1)\alpha(2)$

Each of the four spin functions above might be chosen as possible spin functions for a helium atom in an excited state. However, the last two possibilities are invalid because these functions indicate that the two electrons

(A) have the same spin
(B) can be distinguished
(C) are in the same atomic orbital
(D) have spin quantum numbers that are whole numbers
(E) must move as a pair when the atom changes to another state

43. At 24°C and 0.92 atmosphere, the volume occupied by 5.0 grams of carbon dioxide (molecular weight 44) is

(A) 25 liters
(B) 3.0 liters
(C) 1.0 liter
(D) 0.30 liter
(E) 0.25 liter

GO ON TO THE NEXT PAGE.

44. From the equation

$$dG = -SdT + VdP + \sum_{1}^{K} \mu_i dn_i$$

it can be shown that

(A) $\left(\frac{\partial G}{\partial P}\right)_{T,\, n_i} = V$

(B) $\left(\frac{\partial G}{\partial T}\right)_{P,\, V} = -S$

(C) $\left(\frac{\partial G}{\partial P}\right)_{S,\, V} = \mu_i$

(D) $\left(\frac{\partial G}{\partial T}\right)_{P,\, n_i} = V$

(E) $\left(\frac{\partial S}{\partial G}\right)_{P,\, n_i} = T$

45. The molecular-orbital electron configuration for the oxygen molecule is given by the following expression:

$$(\sigma_{1s})^2 (\sigma_{1s}^*)^2 (\sigma_{2s})^2 (\sigma_{2s}^*)^2 (\sigma_{2p})^2 (\pi_{2p})^4 (\pi_{2px}^*)^1 (\pi_{2py}^*)^1$$

From the electron configuration, it can be determined that the oxygen molecule possesses all of the following properties EXCEPT

(A) bond order = 2
(B) 10 electrons in bonding orbitals
(C) 6 electrons in antibonding orbitals
(D) diamagnetism
(E) ground-state spin multiplicity of 3

46. Which of the following techniques, commonly used by organic chemists, gives information LEAST likely to be useful in elucidating the structure of a newly discovered natural product?

(A) Infrared spectroscopy
(B) Chromatography
(C) Nuclear magnetic resonance spectroscopy
(D) Optical rotatory dispersion
(E) Microanalysis

47. Which of the following compounds has the highest enol content?

(A) $CH_3C(=O)CH_2C(=O)CH_3$

(B) $CH_3C(=O)CH_2CH_2CHO$

(C) $CH_3CH_2CH_2C(=O)CH_3$

(D) $CH_3CH_2CH_2C(=O)CH_2CH_3$

(E) $CH_3CH_2CH_2CH_2CHO$

48. $CH_3CH_2C(=O){-}OH + CH_3CH_2OH \xrightarrow[\text{reflux}]{H^+}$

Which of the following indicates the major products of the reaction above?

(A) $CH_3CH_2C(=O){-}OCH_2CH_2OH + H_2$

(B) $CH_3CH_2{-}C(OH)$ bonded to two O atoms linked by $CH_2{-}CH_2$ (ring) $+ H_2$

(C) $CH_3CH_2C(=O){-}CH(CH_3){-}OH + H_2O$

(D) $CH_3CH_2C(=O){-}CH_2CH_2OH + H_2O$

(E) $CH_3CH_2C(=O){-}OCH_2CH_3 + H_2O$

GO ON TO THE NEXT PAGE.

49. Which of the following alcohols is most difficult to oxidize with an acidic solution of CrO_3?

(A)

$$\begin{array}{l} CH_3CHCH_2CH_2CH_3 \\ \quad\;\; | \\ \quad\; OH \end{array}$$

(B)

$$\begin{array}{c} OH \\ | \\ CH_3-C-CH_2CH_3 \\ | \\ CH_3 \end{array}$$

(C)

$$\begin{array}{l} CH_3CHCH_2CH_2OH \\ \quad\;\; | \\ \quad\; CH_3 \end{array}$$

(D)

$$\begin{array}{l} CH_3CHCH_2CH_3 \\ \quad\;\; | \\ \quad\; CH_2OH \end{array}$$

(E)

$$\begin{array}{l} \qquad\quad OH \\ \qquad\quad\; | \\ CH_3CHCHCH_3 \\ \quad\;\; | \\ \quad\; CH_3 \end{array}$$

50.

$$CH_3\overset{\overset{\displaystyle O}{\|}}{C}CH_3 + C_6H_5-NHNH_2 \xrightarrow[\text{ethanol/water}]{NaOAc + HOAc}$$

Which of the following is the major organic product of the reaction above?

(A) 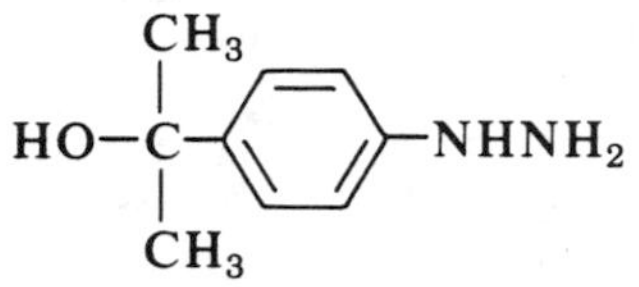 $HO-C(CH_3)_2-C_6H_4-NHNH_2$

(B) $CH_2=C(CH_3)-C_6H_4-NHNH_2$

(C) $C_6H_4(NHNH_2)(C(CH_3)=CH_2)$

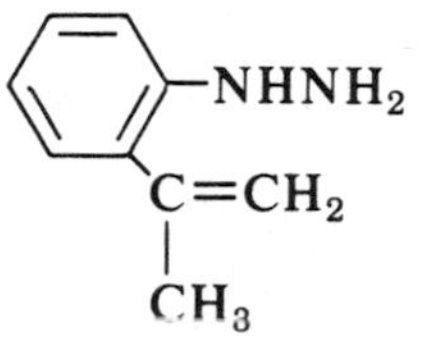

(D) 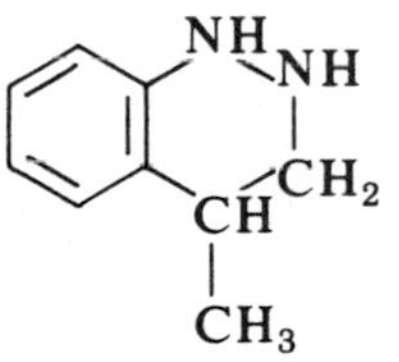 NH, NH, CH_2, CH, CH_3 (ring fused to benzene)

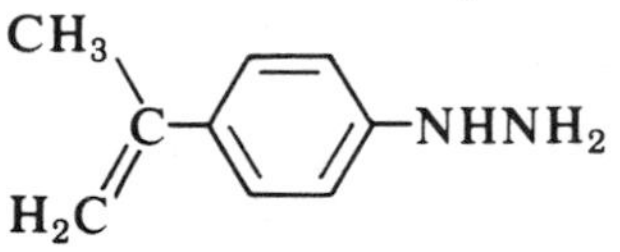

(E) 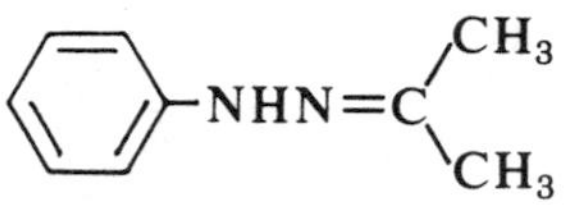 $C_6H_5-NHN=C(CH_3)_2$

GO ON TO THE NEXT PAGE.

51. Of the following, the substance with the highest melting point is

(A) fluorine
(B) dioxygen difluoride
(C) calcium fluoride
(D) silicon tetrafluoride
(E) phosphorus pentafluoride

52. Gases would be produced in all of the following reactions EXCEPT

(A) $NaH + H_2O \rightarrow$

(B) $Na_2CO_3 + HCl \rightarrow$

(C) $Zn(s) + HCl \rightarrow$

(D) $NaCN + HCl \rightarrow$

(E) $Al(OH)_4^- + HCl \rightarrow$

53.

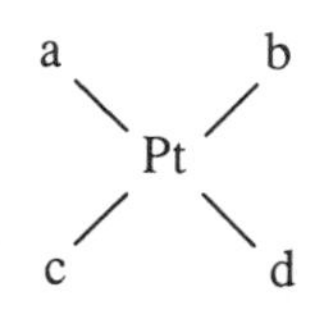

How many isomers can the square planar complex sketched above have?

(A) 0 (B) 2 (C) 3 (D) 4 (E) 6

54. Which of the following is the actual ground state electron configuration of the chromium atom? (Atomic number: Cr = 24)

(A) [Ar] $3d^4 4s^2$

(B) [Ar] $3d^5 4s^1$

(C) [Ar] $3d^6$

(D) [Ar] $3p^6 4s^2$

(E) [Ar] $3p^6$

55. Which of the following molecules can normally behave as an acid according to any one of the three acid definitions, Arrhenius, Brönsted-Lowry, or Lewis?

I. ClOH
II. CsOH
III. SO_3
IV. BH_3

(A) I and III only
(B) I and IV only
(C) I, II, and IV only
(D) I, III, and IV only
(E) II, III, and IV only

56. A pH meter measures

(A) voltage
(B) current
(C) resistance
(D) power
(E) conductance

57. A 60.-milliliter sample of a 0.020-molar solution of Fe^{2+} is titrated with a 0.010-molar solution of $K_2Cr_2O_7$ in acid solution. What volume of the $K_2Cr_2O_7$ solution must be added to reach the equivalence point?

(A) 10. mL
(B) 20. mL
(C) 30. mL
(D) 60. mL
(E) 120 mL

GO ON TO THE NEXT PAGE.

58. All of the following are effective in reducing the amount of time for a controlled-potential electrolysis to reach a fixed concentration EXCEPT

(A) reducing the volume of the sample
(B) reducing the concentration of the species being determined while the volume of the solution is held constant
(C) using electrodes of larger areas
(D) using more efficient stirring
(E) reducing the current as the electrolysis proceeds

59. What is the principal reason for the fact that acetic acid is a suitable solvent for the titrations of weak bases using perchloric acid as titrant?

(A) Acetic acid is a stronger acid than water and levels weak bases placed into it.
(B) Acetic acid is a solvent of low dielectric constant and hence promotes the transfer of a proton from perchloric acid to the base.
(C) Acetic acid is a weaker base than water and competes less effectively for the proton on the protonated base than does water.
(D) Acetic acid as a solvent is a "super acid."
(E) The activity coefficient of the proton is higher in water than in acetic acid.

60. To maximize the reliability of optical absorbance measurements used for calculating concentrations of an absorbing solute, samples of solutions should be prepared that exhibit percent-transmittance values of

(A) less than 1%
(B) 1-10%
(C) 30-50%
(D) 90-99%
(E) greater than 99%

61. Which of the following elements is a semi-conductor?

(A) S
(B) Sc
(C) Si
(D) Sm
(E) Sr

GO ON TO THE NEXT PAGE.

62. $3\,A + B \rightarrow 2\,P$

The following data about initial rates were obtained for the stoichiometric reaction above.

Experiment	$[A]_0$, M	$[B]_0$, M	Initial rate $= d[A]/dt$
1	0.20	0.20	$-1.2 \times 10^{-8}\ M\,s^{-1}$
2	0.20	0.60	$-1.2 \times 10^{-8}\ M\,s^{-1}$

For a third experiment, a plot of $1/[A]$ *versus* time was found to be linear. What is the order of reaction with respect to the concentrations of A and B ?

	Order with Respect to [A]	Order with Respect to [B]
(A)	3	0
(B)	2	1
(C)	2	0
(D)	1	1
(E)	0	2

---

63. The determination from calorimetric data of the absolute entropy of a vapor in equilibrium with liquid at room temperature requires all of the following factors EXCEPT

(A) the assumption that the absolute entropy is zero at the absolute zero of temperature
(B) the heat capacity of the solid as a function of temperature
(C) a means for extrapolating heat capacity data to absolute zero
(D) the value of the partition function for the molecules of vapor
(E) the value for the enthalpy of vaporization of the liquid

64. At 25°C, which of the following substances has the lowest molar entropy?

(A) $N_2(g)$ (B) $Cl_2(g)$ (C) $Mg(s)$
(D) $C_6H_6(\ell)$ (E) $CCl_4(\ell)$

65. At 25°C, which of the following substances has the highest surface tension?

(A) $C_6H_6(\ell)$ (B) $C_6H_{14}(\ell)$
(C) $C_2H_5OH(\ell)$ (D) $H_2O(\ell)$
(E) $C_2H_5-O-C_2H_5(\ell)$

GO ON TO THE NEXT PAGE.

66. $CH_3-CH_2-CH_2-OH$ (I) $CH_3-CH_2-SO_3H$ (II) FCH_2-CH_2-COOH (III)

$CH_3-CHF-COOH$ (IV) $H_3C-C_6H_4-OH$ (V)

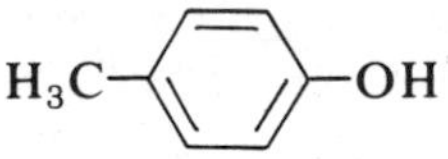

Which of the following indicates the order of decreasing Brönsted-Lowry acid strength, from most acidic to least acidic, for the compounds above?

(A) I > II > III > IV > V
(B) II > IV > III > V > I
(C) II > V > IV > III > I
(D) IV > III > II > I > V
(E) IV > III > II > V > I

---

67. If a mixture of the following compounds is adsorbed at the top of an alumina column and then eluted with solvents, which is most tightly held and thus forms the top band of the chromatogram?

(A)  cyclohexene (B)  cyclohexanone (C) 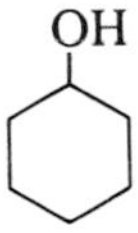 cyclohexanol (D)  benzene (E) 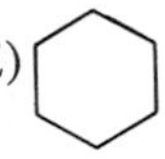 cyclohexane

68. If basicity is defined as tendency to combine with a proton, the LEAST basic of the following compounds in solution is

(A) NH_3

(B) CH_3NH_2

(C) $(CH_3)_3N$

(D) $C_6H_5NH_2$

(E)  $O_2N-C_6H_4-NH_2$

69. OCH_3-benzene-CH_2CH_3 (para) $\xrightarrow[AlCl_3]{Cl_2}$

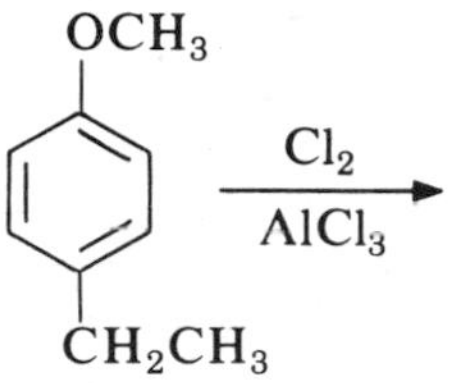

Which of the following is the major product of the reaction above?

(A) 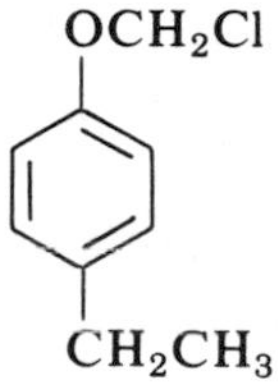 OCH_2Cl / CH_2CH_3 (para-substituted benzene)

(B) 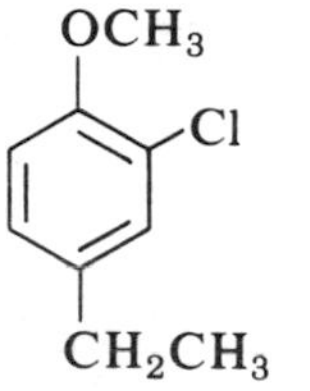 OCH_3, Cl (ortho to OCH_3) / CH_2CH_3

(C) 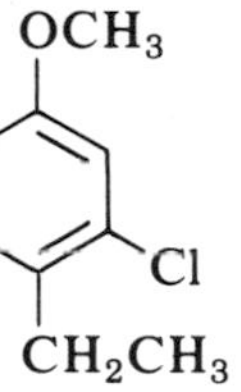 OCH_3 / Cl (ortho to CH_2CH_3), CH_2CH_3

(D) 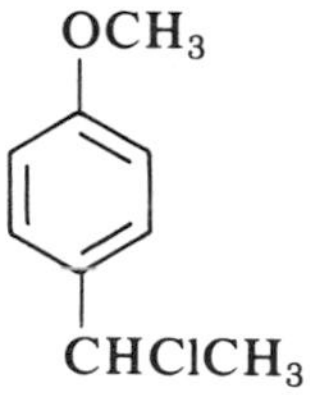 OCH_3 / $CHClCH_3$

(E) 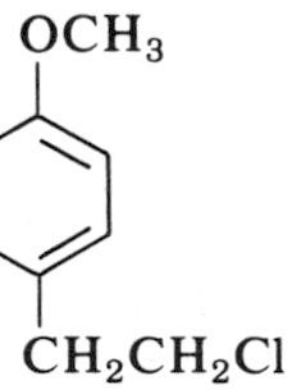 OCH_3 / CH_2CH_2Cl

GO ON TO THE NEXT PAGE.

70. Which of the following is a set of reactants that produces *t*-butyl methyl ether, $(CH_3)_3C—O—CH_3$, in good yield?

(A) $(CH_3)_3C—ONa + CH_3Cl \xrightarrow{(CH_3)_3C—OH}$

(B) $CH_3CO_2C(CH_3)_3 \xrightarrow[\text{2. } H_3O^+]{\text{1. } CH_3MgBr}$

(C) $(CH_3)_3CCl + KOCH_3 \xrightarrow{CH_3OH}$

(D) $(CH_3)_3CCH_3 + O_2 \xrightarrow[370°]{V_2O_5/Al_2O_3}$

(E) $CH_3COC(CH_3)_3 \xrightarrow{H_2/Pd}$

71. In water which of the following is the strongest acid?

(A) HCl (B) HIO (C) HClO
(D) HCN (E) H_2S

72. Element X has an outer electron configuration of $3s^2$. Element Y is directly below element X in the periodic table. A compound having the formula XY would be most likely to have what type of bonding?

(A) Covalent
(B) Ionic
(C) *d-f* overlap
(D) van der Waals forces
(E) Metallic

73. Which of the following carbonyl complexes is predicted to be unstable by the effective atomic number rule? (Atomic numbers: V = 23, Cr = 24, Fe = 26, Co = 27, Ni = 28)

(A) $V(CO)_6$ (B) $Cr(CO)_6$ (C) $Fe(CO)_5$
(D) $Co_2(CO)_8$ (E) $Ni(CO)_4$

74. Which of the following statements concerning the three ions Na^+, Mg^{2+}, and Al^{3+} is correct?

(A) Al^{3+} has the smallest tendency to form complex ions.
(B) Al^{3+} forms the only water-insoluble chloride.
(C) Mg^{2+} forms the least basic hydroxide.
(D) Mg^{2+} forms the most soluble hydroxide.
(E) Na^+ has the largest radius.

75. Which of the following is a *p*-type semiconductor? (Atomic numbers: Ga = 31, Ge = 32, As = 33)

(A) Gallium
(B) Germanium doped with gallium
(C) Germanium
(D) Germanium doped with arsenic
(E) Arsenic

76. An aldol condensation is generally catalyzed by

(A) sodium hydroxide
(B) anhydrous aluminum chloride
(C) mercuric sulfate
(D) ammoniacal cuprous chloride
(E) finely divided nickel

77. Which of the following transformations can be accomplished by the use of thionyl chloride, $SOCl_2$?

(A) $C_6H_5CH{=}CH_2$ to $C_6H_5—\underset{\displaystyle Cl}{\underset{|}{CH}}—CH_2Cl$

(B) $C_6H_5—\underset{\displaystyle OH}{\underset{|}{CH}}—CH_3$ to $C_6H_5—\underset{\displaystyle Cl}{\underset{|}{CH}}—CH_3$

(C) $C_6H_5CH_3$ to $C_6H_5CHCl_2$

(D) $C_6H_5CH_3$ to $C_6H_5\overset{\displaystyle O}{\overset{||}{C}}Cl$

(E) $C_6H_5CH_3$ to $C_6H_5CH_2\overset{\displaystyle O}{\overset{||}{S}}Cl$

GO ON TO THE NEXT PAGE.

78. The addition of cinnamaldehyde, $C_6H_5CH{=}CH{-}CHO$, to a solution of sodium borohydride in ether gives a mixture of cinnamyl alcohol and 3-phenyl-1-propanol. The yield of cinnamyl alcohol can best be increased by

(A) heating the reaction mixture
(B) using an excess of sodium borohydride
(C) stirring more rapidly
(D) using a longer reaction time
(E) adding the hydride to the aldehyde

---

79. Which of the following carbonyl-containing compounds reacts fastest with the nucleophile OH^- ?

(A) Ethyl benzoate
(B) Benzaldehyde
(C) Benzoyl chloride
(D) Benzamide
(E) Sodium benzoate

80. Acetophenone, $C_6H_5COCH_3$, can be isolated in good yield at room temperature by using which of the following procedures with appropriate workup?

(A) $C_6H_5COCl + CH_3Li \longrightarrow$

(B) $C_6H_5COCl + CH_4 \xrightarrow[(C_2H_5)_2O]{BF_3}$

(C) $C_6H_5MgBr + CH_3CO_2CH_3 \longrightarrow$

(D) $C_6H_5Li + CH_3CO_2H \longrightarrow$

(E) $C_6H_6 + CH_3COCl \xrightarrow{AlCl_3}$

81. What is the osmotic pressure of a 0.100-molar aqueous solution of sucrose (molecular weight = 342) at 27°C ? (R = $0.0821\ L \cdot atm \cdot mol^{-1} \cdot K^{-1}$ = $8.314\ J \cdot mol^{-1} \cdot K^{-1}$ = $1.98\ cal \cdot mol^{-1} \cdot K^{-1}$)

(A) 249 atm (B) 125 atm (C) 60.0 atm
(D) 5.40 atm (E) 2.46 atm

82. A reaction proceeds by a two-step mechanism:

$$A_2 \underset{k_{-1}}{\overset{k_1}{\rightleftharpoons}} 2\,A \qquad \text{fast reaction}$$

$$A + B \xrightarrow{k_2} \text{products} \qquad \text{slow reaction}$$

What is the rate law for the overall reaction?

(A) Rate $= k\,[A_2]$

(B) Rate $= k\,[A_2][B]$

(C) Rate $= k\,[A_2]^2[B]$

(D) Rate $= k\,[A_2]^{\frac{1}{2}}[B]$

(E) Rate $= k\,[A_2]^{\frac{1}{2}}[B]^2$

GO ON TO THE NEXT PAGE.

Questions 83-85

Dimethyl ether decomposes in the gas phase according to the equation:

$$(CH_3)_2O(g) \rightarrow CH_4(g) + H_2(g) + CO(g)$$

Experimental measurements at 504°C of the total pressure as a function of time are shown in columns 1 and 2 of the table below. (P_{incr} = increase in pressure; P_{unde} = pressure due to undecomposed ether.)

1 t(secs)	2 P_{total} (mmHg)	3 P_{incr} (mmHg)	4 P_{unde} (mmHg)
0	312	0	312
390	408	96	264
777	498	186	219
1,195	562	250	187
∞	931	619	0

83. If P_{decr} is defined as the decrease in partial pressure of $(CH_3)_2O$ caused by decomposition, which of the following represents the increase in total pressure shown in column 3 of the table? (P_0 = 312 millimeters)

    (A) P_{decr} (B) $2P_{decr}$ (C) $3P_{decr}$
    (D) $P_0 - P_{decr}$ (E) $P_0 + 3P_{decr}$

84. The partial pressure of undecomposed $(CH_3)_2O$ shown as P_{unde} in column 4 of the table represents

    (A) P_{decr} (B) $2P_{decr}$ (C) $3P_{decr}$
    (D) $P_0 - P_{decr}$ (E) $P_0 + 3P_{decr}$

85.

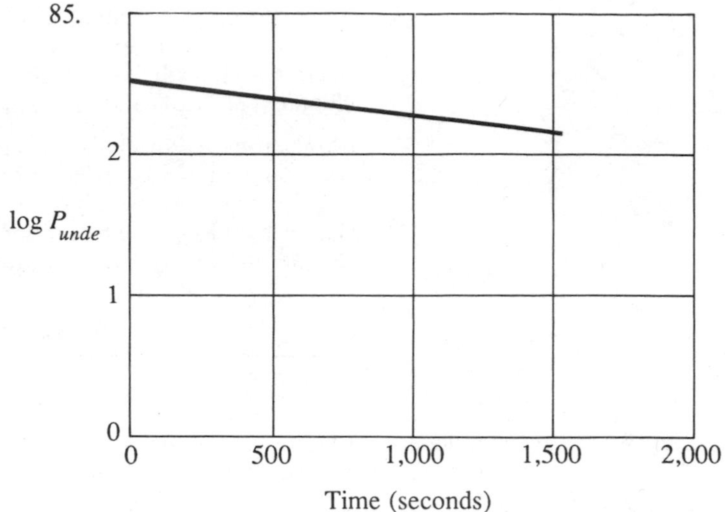

A plot of $\log P_{unde}$ against time gives a straight line, as shown above. One can conclude, therefore, that the decomposition of $(CH_3)_2O$ in the gas phase at 504°C is

(A) a zero-order reaction
(B) a first-order reaction
(C) a photochemical reaction
(D) a chain reaction
(E) an explosive reaction

GO ON TO THE NEXT PAGE.

86. Oxalic acid can be determined gravimetrically by precipitation as $Ce_2(C_2O_4)_3 \cdot XH_2O$ and ignition to CeO_2 for weighing. The gravimetric factor for conversion of CeO_2 (formula weight 172.1) to $H_2C_2O_4$ (formula weight 90.03) is

(A) $\frac{(90.03)}{(172.1)}$

(B) $\frac{(172.1)}{(90.03)}$

(C) $\frac{2(172.1)}{3(90.03)}$

(D) $\frac{3(90.03)}{2(172.1)}$

(E) $\frac{2(90.03)}{3(172.1)}$

87.

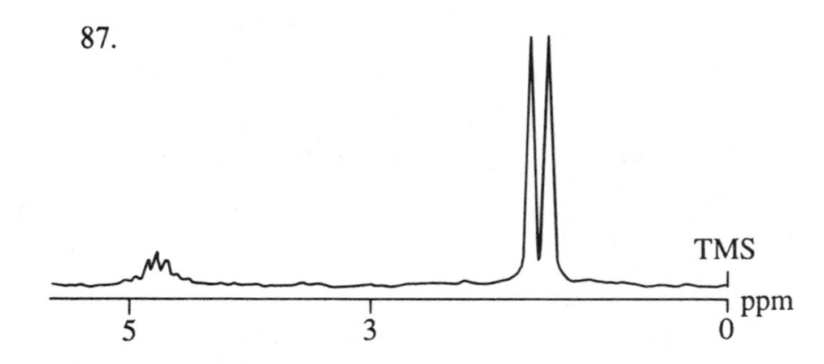

The nuclear magnetic resonance spectrum shown above is the spectrum for which of the following compounds?

(A) $CH_3—CH_2—CH_2—NO_2$

(B) $(CH_3)_2CH—NO_2$

(C) $(CH_3)_2C(OH)—NO_2$

(D) $CH_3CH_2—NO_2$

(E) $CH_3(CH_2)_7NO_2$

88. Which of the following pieces of information should be used to determine by the Q test whether a suspect number in a series of five replicate measurements should be rejected?

I. The mean of the five values
II. The median of the five values
III. The range of the five values
IV. The difference between the suspect number and its nearest neighbor

(A) I and II only
(B) I and III only
(C) II and III only
(D) III and IV only
(E) I, III, and IV

89. An indicator has K_a equal to 8×10^{-7}. What is the ratio of the concentration of the acid form of the indicator to the concentration of the basic form, ($[HIn]/[In^-]$), in a buffer solution in which $[H^+]$ is 1×10^{-6}-molar ?

(A) $\frac{1}{80}$ (B) $\frac{1}{8}$ (C) $\frac{0.8}{1}$ (D) $\frac{1}{0.8}$ (E) $\frac{8}{1}$

GO ON TO THE NEXT PAGE.

90.

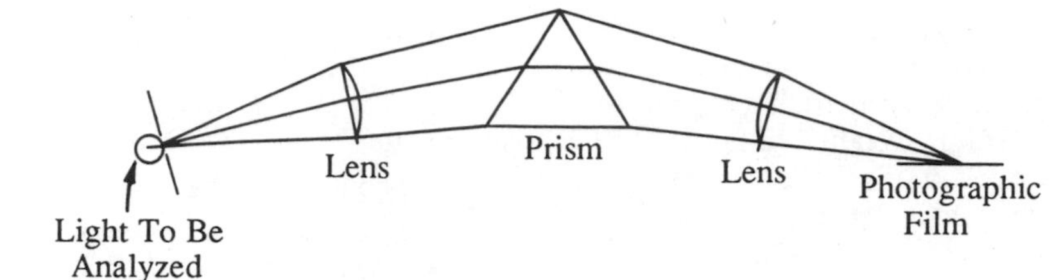

The schematic diagram above represents what type of instrument?

(A) A nephelometer
(B) A spectrograph
(C) A spectrophotometer
(D) A refractometer
(E) A polarimeter

---

91. Which of the following statements is true of the element with atomic number 55 and atomic weight 133 ?

(A) An atom of the element contains only 78 nucleons.
(B) The fourth major energy level, or N shell, contains 32 electrons.
(C) Atoms of the element possess no $4f$ electrons.
(D) Atoms of the element contain no partially filled subshells.
(E) The element is an inert gas.

92. If the mean free path of a gaseous molecule is 50. centimeters at a pressure of 1.0×10^{-4} millimeter mercury, what is its mean free path when the pressure is increased to 1.0×10^{-2} millimeter mercury?

(A) 5.0×10^{-1} cm
(B) 5.0 cm
(C) 5.0×10^{1} cm
(D) 5.0×10^{2} cm
(E) 5.0×10^{3} cm

93. The limiting molar conductivities, $\Lambda°$ in siemens centimeters^2 mole^{-1}, appropriate where ions move freely through a solution, for certain compounds are as follows: NaCl = 126, KBr = 152, KCl = 150. It can be calculated that $\Lambda°$ for NaBr is

(A) 128 S cm^2/mole
(B) 176 S cm^2/mole
(C) 224 S cm^2/mole
(D) 276 S cm^2/mole
(E) 302 S cm^2/mole

94. $H_2C{=}C{=}CH_2$ (allene structure)

Which of the following is the only symmetry element that the molecule allene shown above (point group D_{2d}) does NOT possess?

(A) C_2 (twofold axis of symmetry)
(B) i (center of symmetry)
(C) σ (plane of symmetry)
(D) S_4 (fourfold rotation-reflection axis)
(E) E (identity operation)

GO ON TO THE NEXT PAGE.

95. 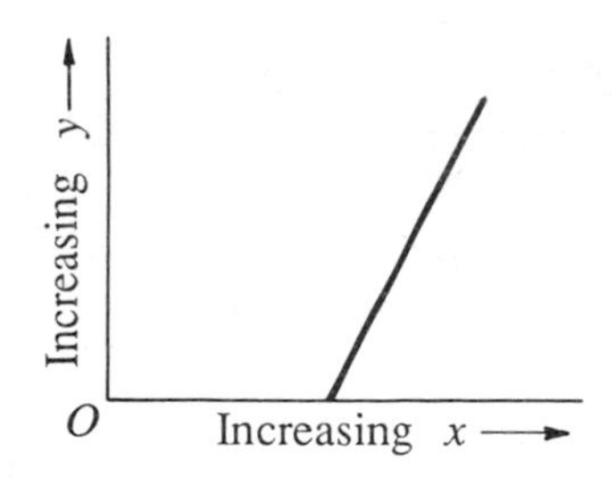

A graph of y plotted against x is a straight line which has the slope shown above and which does not pass through the origin. This could be a graph in which y is the

(A) volume of an ideal gas and x is the absolute temperature
(B) pressure of an ideal gas and x is the reciprocal volume
(C) specific heat of a solid element and x is the absolute temperature
(D) energy of the most energetic electrons emitted during illumination and x is the frequency of the light
(E) first ionization energy and x is the atomic number

96. Which of the numbered hydrogen atoms in the compound below would be most susceptible to free-radical halogenation?

CH_3(1)–C_6H_4–$CH(CH_3)$(2; H 3)–CH(H 4)$=CH$–C_6H_4–H(5)

(A) 1 (B) 2 (C) 3 (D) 4 (E) 5

---

97. Which of the following reactions does NOT involve the formation of a reactive cationic intermediate?

(A) $CH_3–C_6H_5 + Br_2 \xrightarrow{\text{heat}} Br–CH_2–C_6H_5$

(B) $CH_3–C(CH_3)_2–O–C(CH_3)_2–CH_3 + HI \xrightarrow{\text{heat}} 2CH_3–C(CH_3)_2–I$

(C) $CH_3–C(CH_3)_2–OH + HBr \longrightarrow CH_3–C(CH_3)_2–Br$

(D) $C_6H_6 + CH_3Cl \xrightarrow{AlCl_3} C_6H_5–CH_3$

(E) $CH_3–CH=CH–CH_3 + Br_2 \xrightarrow[25°]{CCl_4} CH_3–CHBr–CHBr–CH_3$

GO ON TO THE NEXT PAGE.

98. The resonance approach is an important method for describing the electronic structures of organic molecules. Which of the following is NOT an appropriate use of resonance symbolism?

(A) $CH_3{-}C({=}O){-}O^{\ominus} \longleftrightarrow CH_3{-}C(-O^{\ominus}){=}O$

(B) (benzene Kekulé structure) $\longleftrightarrow$ (benzene Kekulé structure)

(C) $CH_3CH{=}CH\overset{\oplus}{C}HCH_3 \longleftrightarrow CH_3\overset{\oplus}{C}HCH{=}CHCH_3$

(D) $\oplus$(cyclohexadiene)$=CH_2 \longleftrightarrow$ (benzene)$-\overset{\oplus}{C}H_2$

(E) (benzene ring with $\oplus$) $\longleftrightarrow$ (benzene ring with $\oplus$)

99. Which of the following pairs of compounds in ether solution could be most easily separated by extraction with dilute sodium bicarbonate?

(A) CH_3CO_2H and CCl_3CO_2H

(B) $C_6H_5{-}OH$ and $C_6H_5{-}\overset{O}{\overset{\|}{C}}{-}OH$

(C) $C_6H_5{-}CH_2OH$ and $O_2N{-}C_6H_4{-}CH_2OH$

(D) $C_6H_5{-}CO_2H$ and $C_6H_5{-}SO_3H$

(E) CH_3NO_2 and $CH_2(CO_2C_2H_5)_2$

GO ON TO THE NEXT PAGE.

100.

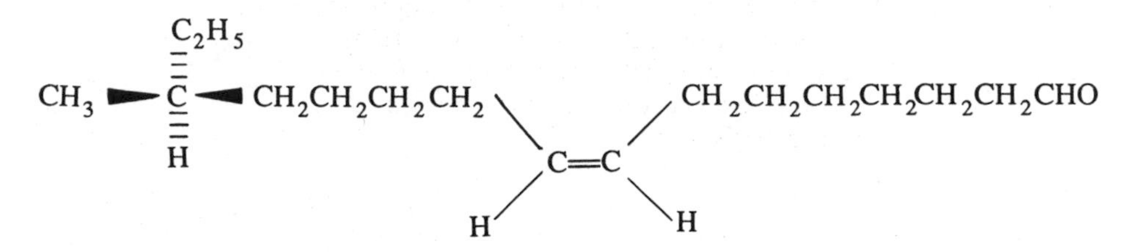

The compound above has recently been identified as a sex pheromone for a particular beetle. The proper designation for the chiral center and the geometrical configuration, respectively, of this molecule would be

(A) *R* and *Z*
(B) *R* and *E*
(C) *S* and *Z*
(D) *S* and *E*
(E) *S* and *trans*

---

101. The type of chemical bonding responsible for the inert gas hydrates, such as $Rn{\cdot}6\ H_2O$, is best described as

(A) covalent bonding
(B) coordinate bonding
(C) ionic attraction
(D) dipole-dipole attraction
(E) dipole-induced dipole attraction

102. Which of the following statements does NOT describe correctly the space lattice with which it is paired?

(A) 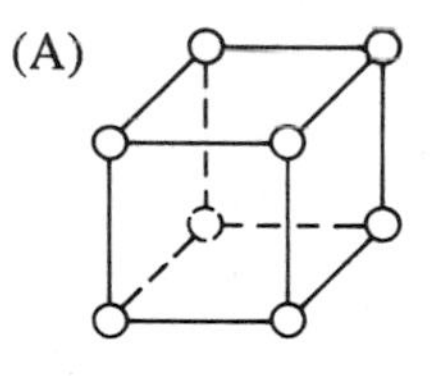 In a substance that crystallizes in the simple cubic form, each atom has six nearest neighbors.

(B) 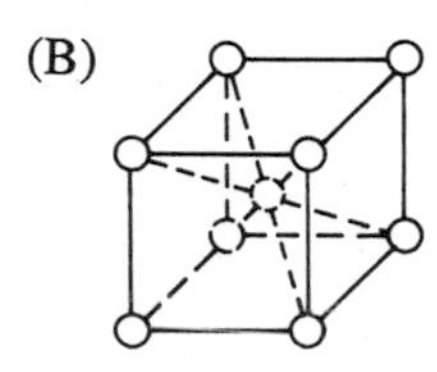 A unit cell of this type contains the equivalent of two particles.

(C) 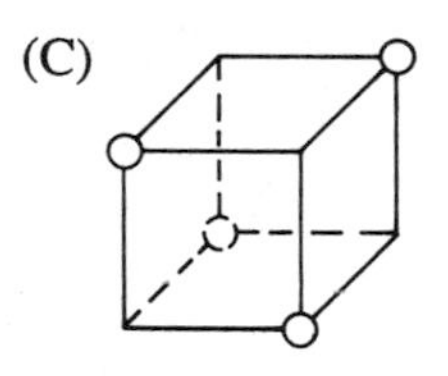 A unit cell of this type contains the equivalent of one particle.

(D) 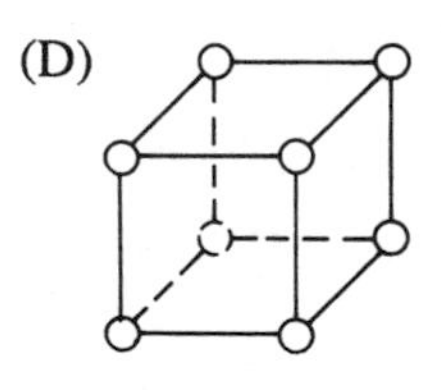 A unit cell of this type contains the equivalent of one particle.

(E) 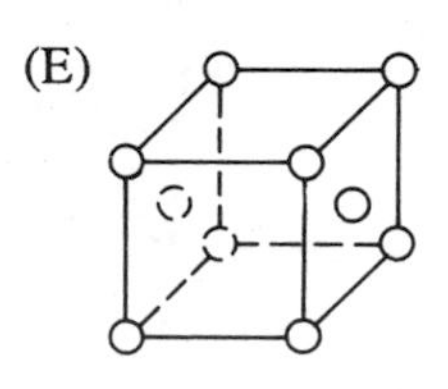 A unit cell of this type contains the equivalent of two particles.

GO ON TO THE NEXT PAGE.

103. Equal numbers of moles of aluminum, magnesium, potassium, sodium, and zinc are treated with excess hydrochloric acid. The greatest quantity of hydrogen is obtained with

(A) Al (atomic weight 27)
(B) Mg (atomic weight 24)
(C) K (atomic weight 39)
(D) Na (atomic weight 23)
(E) Zn (atomic weight 65)

104. Liquid sulfur is very viscous at 200°C because at this temperature

(A) the liquid is made up of long chains of sulfur atoms
(B) individual sulfur atoms are formed
(C) strong bonds are formed among the S_8 ring molecules
(D) liquid sulfur becomes amorphous at this temperature
(E) the rhombic crystalline form converts to the monoclinic form

---

105. Which of the underlined products in the following reactions is NOT an inorganic ring species but is an inorganic cage species?

(A) $3\ S_4N_4 + 6\ Cl_2 \longrightarrow \underline{4\ N_3S_3Cl_3}$

(B) $3\ NH_4Cl + 3\ LiBH_4 \longrightarrow \underline{B_3N_3H_6} + 3\ LiCl + 9\ H_2$

(C) $3\ NH_4Cl + 3\ PCl_5 \longrightarrow \underline{P_3N_3Cl_6} + 12\ HCl$

(D) $2\ BH_4^- + 5\ B_2H_6 \xrightarrow[100°\ C\ -\ 108°\ C]{\text{pyridine}} \underline{B_{12}H_{12}^{2-}} + 13\ H_2$

(E) $6\ S_2Cl_2 \xrightarrow[\Delta]{NH_4Cl} \underline{S_4N_4} + S_8 + 16\ HCl\uparrow$

106. Alkyl halides undergo nucleophilic displacement reactions of the type $RCl + X^- \rightarrow RX + Cl^-$, where X^- may represent a variety of nucleophilic reagents. Among the following compounds, the LEAST reactive halide in nucleophilic displacement reactions carried out under similar conditions is

(A) $CH_2{=}CHCH_2Cl$

(B) $CH_3CH(Cl)CH_2CH_3$

(C) $(CH_3)_3CCl$

(D) $CH_2{=}CH{-}CH_2{-}CH_2Cl$

(E) $CH_3CH_2CH_2CH_2Cl$

GO ON TO THE NEXT PAGE.

107. H_3C CH_3 CH_3 =O

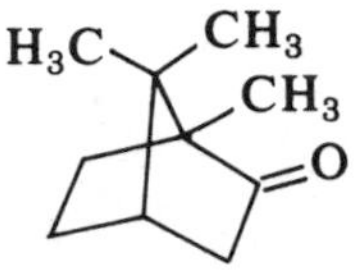

Camphor

Which of the following is the major product of the reaction between camphor, represented above, and lithium aluminum hydride?

(A) H_3C CH_3 CH_3

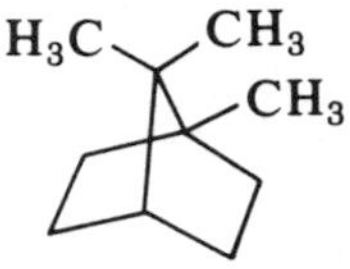

(B) H_3C CH_3 CH_3 OH H

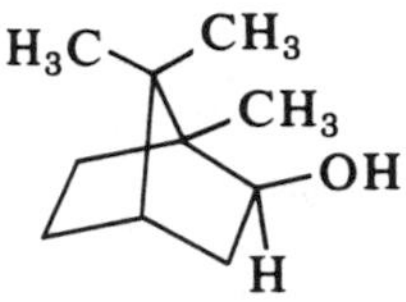

(C) CH_3 $(CH_3)_2HC$ O

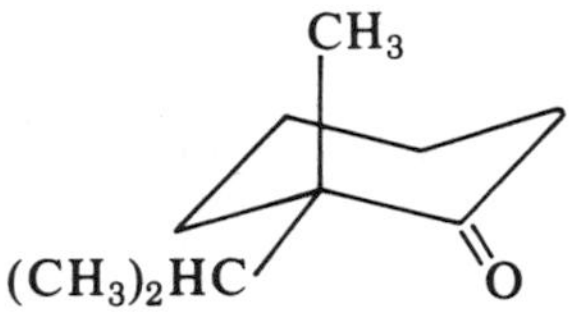

(D) H_3C CH_3 CH_3

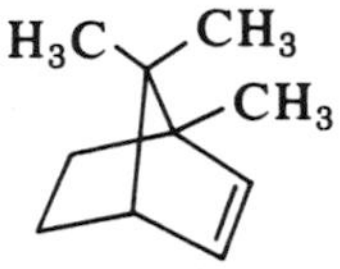

(E) H_3C CH_3 CH_3 =O O

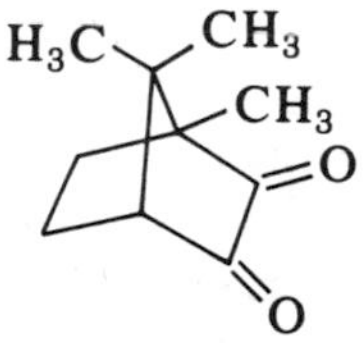

108. Which of the following molecules yields a ketone when it reacts with water, acid, and a catalyst?

(A) $-OSO_3H$

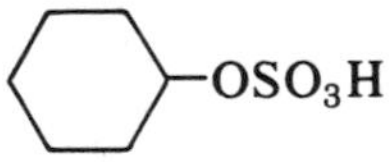

(B)

(C) $-\overset{O}{\overset{\|}{C}}-Cl$

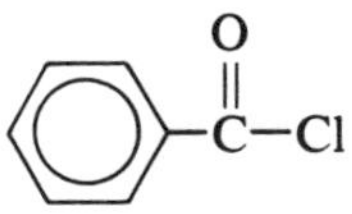

(D) $-C\equiv C-$

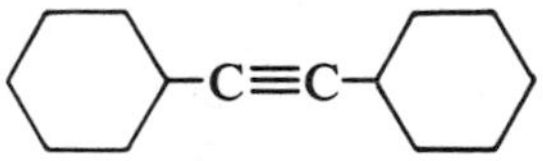

(E) $-\overset{O}{\overset{\|}{C}}-O-\overset{O}{\overset{\|}{C}}-$

GO ON TO THE NEXT PAGE.

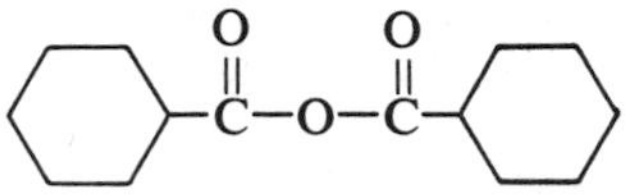

109.

Hückel's rule states that aromaticity exists only for planar cyclic compounds that have $(4n + 2)$ π-electrons ($n = 0$ or any positive integer). According to this rule, which of the structures above must show aromatic properties?

(A) I only
(B) I, III, and IV only
(C) I, II, III, and IV only
(D) II, III, IV, and V only
(E) I, II, III, IV, and V

---

110. Which of the following structural isomers can be resolved into optically active forms?

(A) *trans*-1,3-Cyclohexanediol
(B) *trans*-1,4-Cyclohexanediol
(C) *cis*-1,2-Cyclohexanediol
(D) *cis*-1,3-Cyclohexanediol
(E) *cis*-1,4-Cyclohexanediol

---

111. Which of the following is an example of an organometallic compound participating in an insertion reaction?

(A) $HCo(CO)_4 \longrightarrow H^+ + Co(CO)_4^-$

(B) $Fe(CO)_5 + PPh_3 \longrightarrow Fe(CO)_4PPh_3 + CO$

(C) Me_3SnCl + N(pyridine) $\longrightarrow Me_3SnCl$ N(pyridine)

(D) $MeMn(CO)_5 \longrightarrow Me{-}\overset{\overset{\displaystyle O}{\|}}{C}{-}Mn(CO)_4$

(E) $(HC{=}CH_2)_2Fe(CO)_3$ (butadiene, HC=CH₂ / HC=CH₂ bonded to Fe—CO, CO, CO) $\longrightarrow$ (cyclohexane ring) $Fe(CO)_3$

GO ON TO THE NEXT PAGE.

112. Which of the following types of compounds is most likely to be nonstoichiometric?

(A) Hydrides of transition metals
(B) Halogen compounds of hydrogen
(C) Halogen compounds of nitrogen
(D) Halogen compounds of oxygen
(E) Interhalogen compounds

113. CO coordinates most strongly with metals of zero or low oxidation state primarily because of

(A) the *trans* effect
(B) the intense polarity of the CO molecule
(C) the necessity of following the effective atomic number (EAN) rule
(D) a decrease in the repulsive forces
(E) an increased opportunity for pi bonding

114. The primary reason for the nonexistence of $NaCl_2$ is that

(A) Na has a high second ionization energy
(B) the expected repulsion between the Cl atoms is strong
(C) $NaCl_2$ would have very high lattice energy
(D) the Cl_2^- ion has been observed only as a ligand in coordination complexes of transition metal ions
(E) Cl has a low electron affinity

115.

Ion	P(cm^{-1})	Ligand	Δ_0 (cm^{-1})
Fe^{2+}	17,600	$6H_2O$	10,400
		$6CN^-$	33,000
Co^{3+}	21,000	$6F^-$	13,000

Both of the ions have six $3d$ electrons.

The table above lists values of mean electron-pairing energies P and crystal field splittings Δ_0 with certain ligands. According to these values, which of the following statements is true? (Atomic numbers: Fe = 26, Co = 27)

(A) $Fe(H_2O)_6^{2+}$ and $Fe(CN)_6^{4-}$ are both paramagnetic complexes.
(B) $Fe(CN)_6^{4-}$ is a paramagnetic complex, whereas $Co(F)_6^{3-}$ is a diamagnetic complex.
(C) $Fe(CN)_6^{4-}$ and $Co(F)_6^{3-}$ are both diamagnetic complexes, whereas $Fe(H_2O)_6^{2+}$ is a paramagnetic complex.
(D) $Fe(CN)_6^{4-}$ is a diamagnetic complex, whereas $Fe(H_2O)_6^{2+}$ is a paramagnetic complex.
(E) $Fe(H_2O)_6^{2+}$, $Fe(CN)_6^{4-}$, and $Co(F)_6^{3-}$ are all diamagnetic complexes.

116. The electronegativity of which of the following elements is nearest to that of hydrogen?

(A) Na
(B) Ba
(C) C
(D) O
(E) F

117. What is the wavelength of protons with a mass of 1.66×10^{-27} kilogram and a speed of 2.0×10^5 meters per second? (Planck's constant = 6.62×10^{-34} J · s; speed of light = 2.99×10^8 m/s)

(A) $(6.62 \times 10^{-34})(2.99 \times 10^8)$ m

(B) $\dfrac{(1.66 \times 10^{-27})(2.0 \times 10^5)}{(6.62 \times 10^{-34})(2.99 \times 10^8)}$ m

(C) $\dfrac{(6.62 \times 10^{-34})}{(1.66 \times 10^{-27})(2.99 \times 10^8)}$ m

(D) $\dfrac{(2.0 \times 10^5)(2.99 \times 10^8)}{(6.62 \times 10^{-34})}$ m

(E) $\dfrac{(6.62 \times 10^{-34})}{(1.66 \times 10^{-27})(2.00 \times 10^5)}$ m

118. $$N_2O_4 \rightleftarrows 2\ NO_2$$

Undecomposed N_2O_4 is introduced into an empty flask until the pressure of the pure N_2O_4 is 1 atmosphere. At a certain temperature, 40 percent of the N_2O_4 is decomposed to NO_2 according to the reaction above. Approximately what is the value of the pressure equilibrium constant, K_p, at this temperature?

(A) 10^{-4}
(B) 10^{-2}
(C) 1
(D) 10^2
(E) 10^4

119. The most accurate value for the interatomic distance in a polar diatomic molecule is determined from data given by

(A) infrared spectroscopy
(B) microwave spectroscopy
(C) Raman spectroscopy
(D) ultraviolet spectroscopy
(E) nuclear magnetic resonance spectroscopy

GO ON TO THE NEXT PAGE.

120. The presence of a weak band in the infrared absorption spectrum of HCl gas at a frequency roughly twice that of the strong fundamental band centered at 2,886 centimeters^{-1} can be ascribed to

(A) the presence of different isotopes of chlorine in HCl
(B) transitions from the $v = 1$ to the $v = 2$ vibrational level
(C) transitions from the $v = 0$ to the $v = 2$ vibrational level
(D) changes in electronic energy
(E) changes in rotational energy

121. At a particular wavelength, the molar absorptivity $\mathcal{E}$ of benzoic acid is 2×10^3. When an unknown solution of benzoic acid is measured at this wavelength in a 5-centimeter cell, the percent transmitted light is 20. The molar concentration of benzoic acid in the unknown solution is most nearly which of the following?

(A) $\dfrac{-\log 20}{(2 \times 10^3)(5)}\ M$

(B) $\dfrac{-\log 0.20}{(2 \times 10^3)(5)}\ M$

(C) $\dfrac{-\log 0.80}{(2 \times 10^3)(5)}\ M$

(D) $\dfrac{(2 \times 10^{-3})(5)}{-\log 0.80}\ M$

(E) $\dfrac{(2 \times 10^3)(5)}{-\log 0.20}\ M$

---

122. Hydrochloric acid is titrated with sodium hydroxide, and the conductance of the solution is measured as a function of the volume of NaOH added. Which of the following curves best represents this titration?

(A)

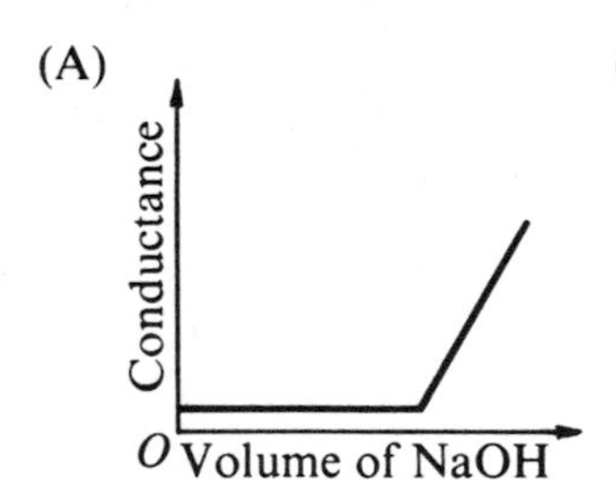

(B)

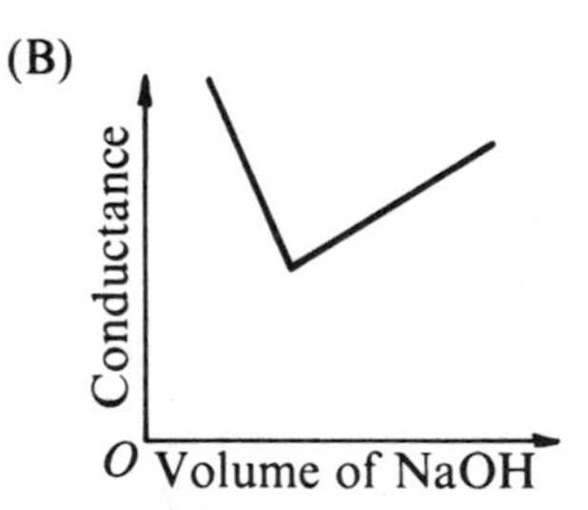

(C)

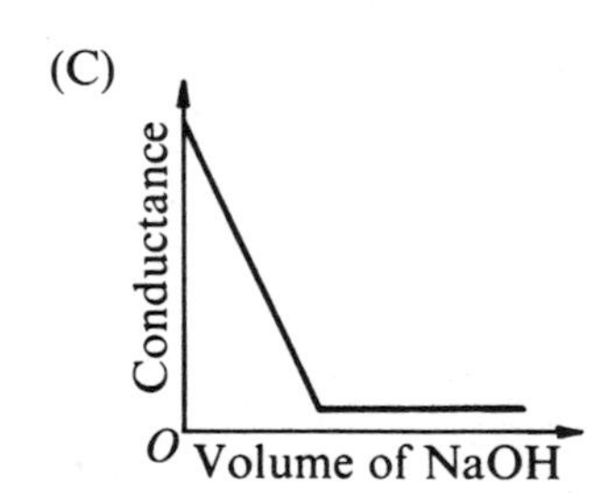

(D)

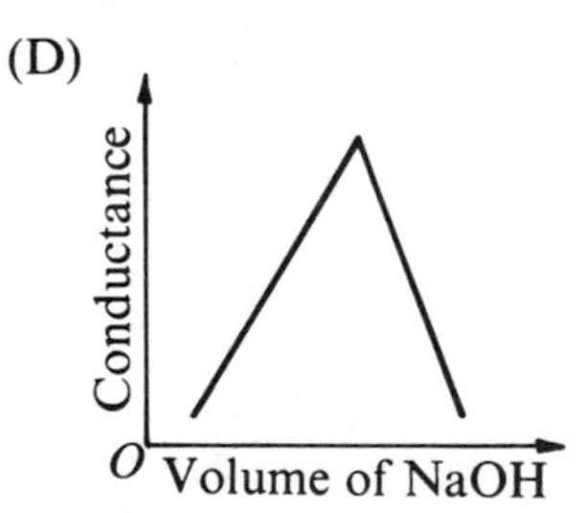

(E)

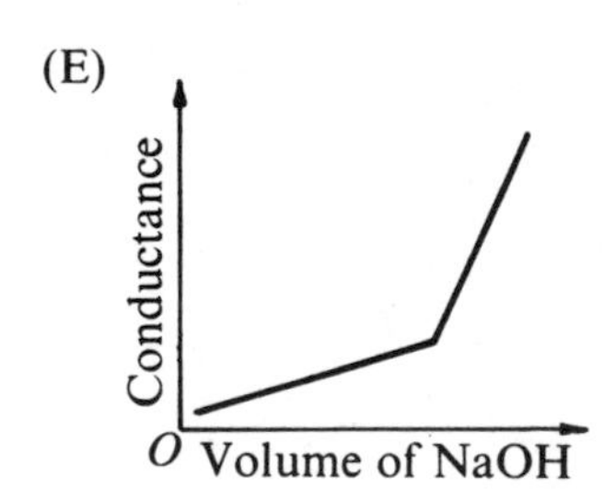

GO ON TO THE NEXT PAGE.

123. Which of the following substances would be best for the standardization of a potassium permanganate solution?

(A) $KBrO_3$

(B) $K_2Cr_2O_7$

(C) $Na_2C_2O_4$

(D) As_2O_5

(E) $Na_2S_2O_3$

124. Two substances A and B have the ultraviolet absorption characteristics shown below:

Substance	Molar Absorptivity at λ_1	Molar Absorptivity at λ_2
A	4,120	0.00
B	3,610	300.

A solution containing both of these substances gives an absorbance of 0.754 at λ_1 in a 1.00-centimeter cell and an absorbance of 0.240 at λ_2 in a 10.00-centimeter cell. What is the concentration of B in mole per liter?

(A) $0.64 \times 10^{-5}\ M$
(B) $0.80 \times 10^{-5}\ M$
(C) $0.64 \times 10^{-4}\ M$
(D) $0.80 \times 10^{-4}\ M$
(E) $1.12 \times 10^{-4}\ M$

125. When the salt of a strong base and a weak acid is dissolved in water, the anion of the salt undergoes hydrolysis and a basic solution results. What is the equilibrium constant for this reaction?
(K_{eq} = equilibrium constant, K_w = ion product of water, K_a = ionization constant of the acid, K_b = ionization constant of the base)

(A) K_w/K_b

(B) K_a/K_w

(C) K_w/K_a

(D) $K_w/[H^+]$

(E) $\sqrt{K_w/K_a}$

---

126. 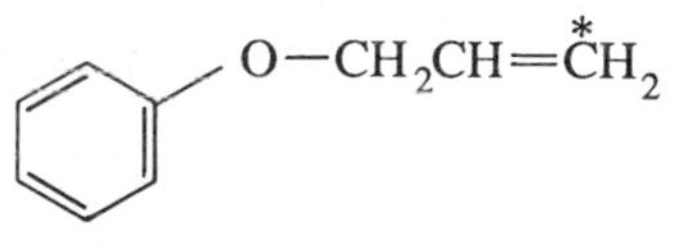 $O{-}CH_2CH{=}\overset{*}{C}H_2$

The Claisen rearrangement of allyl phenyl ether, labeled with C-14 as shown, yields which of the following?

(A) 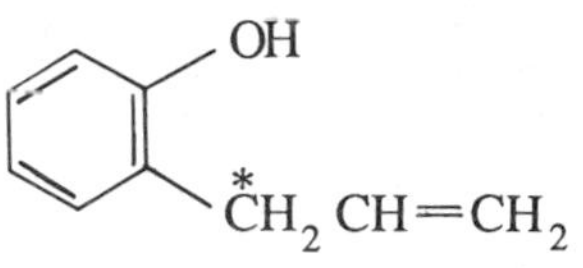 OH; $\overset{*}{C}H_2\ CH{=}CH_2$

(B) O; CH_2; CH_2; $\overset{*}{C}H_2$

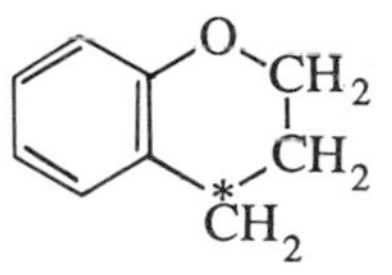

(C) OH; $CH_2CH{=}\overset{*}{C}H_2$

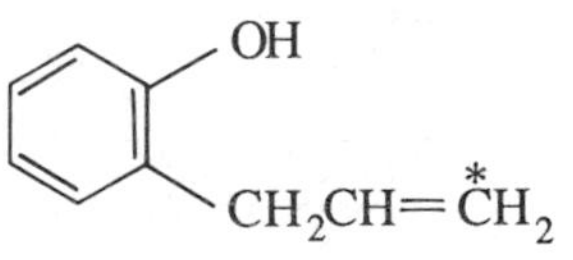

(D) 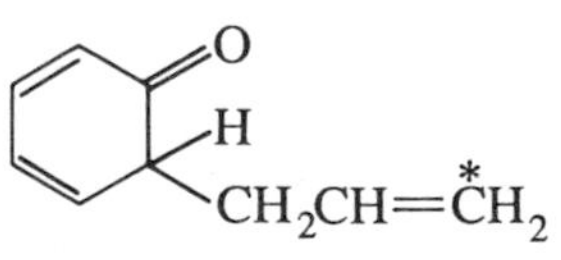 O; H; $CH_2CH{=}\overset{*}{C}H_2$

(E) 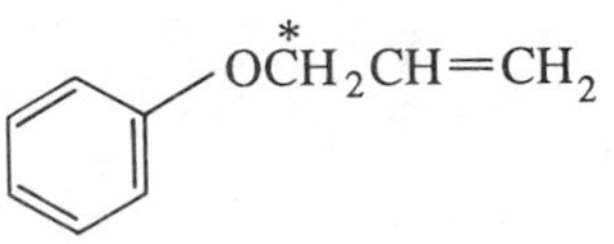 $O\overset{*}{C}H_2CH{=}CH_2$

GO ON TO THE NEXT PAGE.

127. Which of the following transformations can be accomplished with the Wittig reagent, $(C_6H_5)_3P = CH_2$?

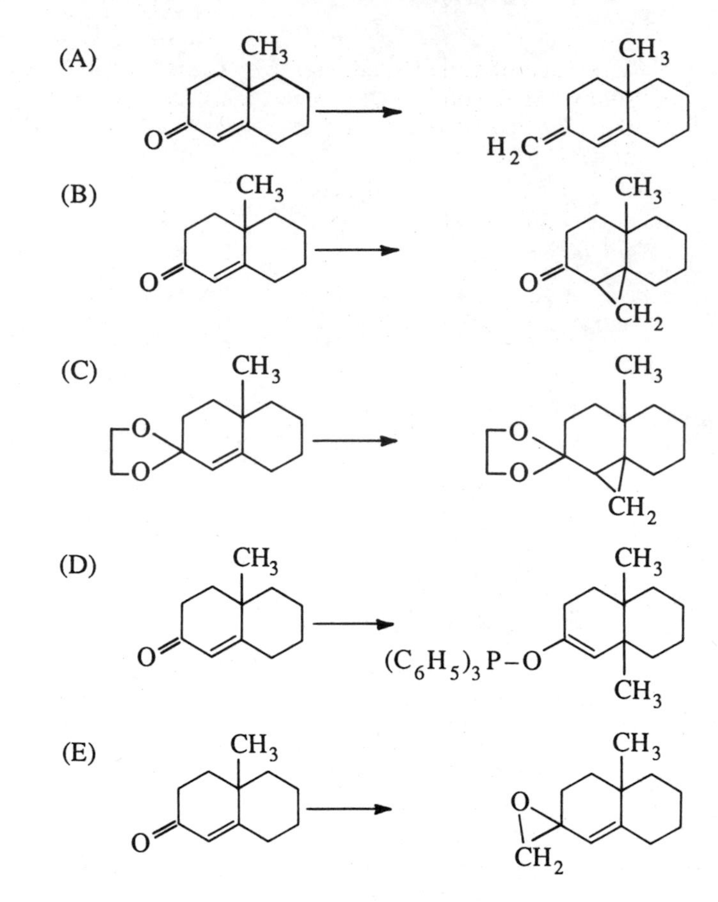

GO ON TO THE NEXT PAGE.

128. $CH_3O_2C{-}(CH_2)_4{-}CO_2CH_3 \xrightarrow[CH_3OH]{NaOCH_3}$

Treatment of dimethyl adipate with methanolic sodium methoxide, as shown above, yields which of the following products after workup with cold dilute aqueous acid?

(A) $(CH_3O)_3C{-}(CH_2)_4{-}C(OCH_3)_3$

(B) 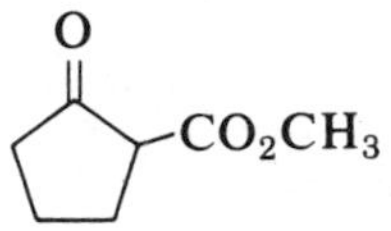 O, CO_2CH_3

(C) 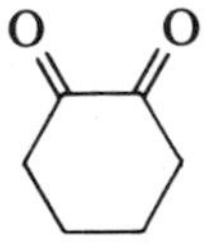 O O

(D) 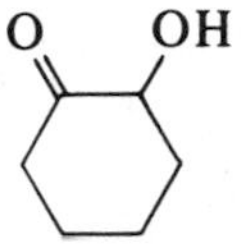 O OH

(E) 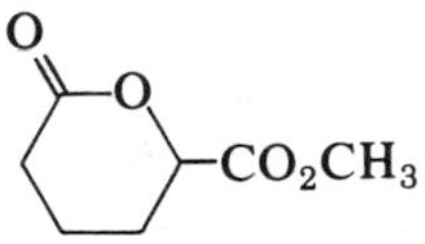 O, O, CO_2CH_3

---

129. Which of the following routes for the synthesis of 1-propanol would NOT proceed as indicated?

(A) $CH_3CH{=}CH_2 \xrightarrow{B_2H_6} (CH_3CH_2CH_2)_3B \xrightarrow[NaOH]{H_2O_2} CH_3CH_2CH_2OH$

(B) $CH_3CH_2CO_2H \xrightarrow[2.\ H_2O]{1.\ LiAlH_4} CH_3CH_2CH_2OH$

(C) $CH_3CH{=}CH_2 \xrightarrow[peroxides]{HBr} CH_3CH_2CH_2Br \xrightarrow{NaOH} CH_3CH_2CH_2OH$

(D) $CH_3CH_2Br \xrightarrow[ether]{Mg} CH_3CH_2MgBr \xrightarrow[2.\ H_2O]{1.\ CH_2O} CH_3CH_2CH_2OH$

(E) $CH_3CH{=}CH_2 \xrightarrow[peroxides]{H_2O,\ H_2SO_4} CH_3CH_2CH_2OH$

GO ON TO THE NEXT PAGE.

130. Which of the following statements best describes the mechanism of the reaction between styrene ($C_6H_5CH=CH_2$) and hydrogen bromide in ether with no peroxides present?

(A) The H and Br atoms add to the double bond in *syn* fashion in a concerted reaction, producing β-bromoethylbenzene.

(B) Br^- attacks the double bond at the terminal carbon in the slow step, yielding an intermediate benzylic carbanion, which then picks up a proton.

(C) Br^- attacks the double bond in the slow step, producing a three-membered cyclic anionic intermediate, which then protonates in *anti* fashion to yield α-bromoethylbenzene.

(D) The double bond is protonated on the terminal carbon in the slow step, yielding an intermediate benzylic carbocation, which then is attacked by bromide.

(E) A Br^+ ion attacks the double bond in the slow step, producing an intermediate three-membered cyclic cation, which then is attacked in *anti* fashion by a hydride ion to form α-bromoethylbenzene.

131.

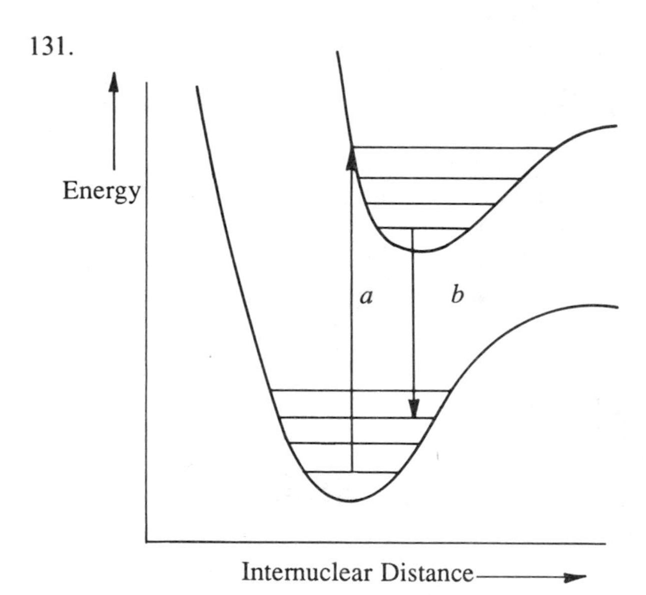

The process indicated by the successive steps *a* and *b* in the diagram above is an example of

(A) fluorescence
(B) phosphorescence
(C) predissociation
(D) radiationless transition
(E) Stark effect

132. The work required to increase the volume of a system by 1.00 liter against an external pressure of 1.00 atmosphere is

(A) 4.18 joules
(B) 8.31 joules
(C) 0.0821 liter-atmosphere
(D) 1.00 liter-atmosphere
(E) 82.0 liter-atmospheres

133. The activated-complex or transition-state theory provides the following equation for the second-order rate constant:

$$k_2 = \frac{kT}{h} e^{\frac{\Delta S^*}{R}} e^{-\frac{\Delta H^*}{RT}}$$

An important assumption in the development of this equation is that the

(A) free energy of activation is negative
(B) entropy of activation is positive
(C) enthalpy of activation (ΔH^*) is independent of temperature
(D) reactants and the activated complex are in equilibrium
(E) activated complex possesses no vibration energy

134. Which of the following substances is an explosive, is a high-nitrogen fertilizer, and, on slow thermal decomposition, yields nitrous oxide?

(A) $CaCN_2$
(B) NH_4NO_3
(C) NH_4Cl
(D) $(NH_4)_2SO_4$
(E) NaOCN

GO ON TO THE NEXT PAGE.

135. $$\frac{N_j}{N} = \frac{g_j \exp(-\mathcal{E}_j/kT)}{\sum_i g_i \exp(-\mathcal{E}_i/kT)}$$

One form of the Boltzmann distribution law is shown above, where

N_j is the number of molecules in the sample occupying the jth quantum level;

g_j is the degeneracy of the jth level;

$\mathcal{E}_j$ is the energy of the jth level;

and N is the total number of molecules in the sample.

The partition function for a molecule in a system obeying this distribution law is defined as

(A) $\frac{N_j}{N}$

(B) $\exp(-\mathcal{E}_j/kT)$

(C) $\mathcal{E}_j/kT$

(D) $g_j \exp(-\mathcal{E}_j/kT)$

(E) $\sum_i g_i \exp(-\mathcal{E}_i/kT)$

136. Which of the following statements about the group of elements consisting of Li, Na, K, Rb, and Cs is correct?

(A) The metals are all powerful oxidizing agents.
(B) They are known as the alkaline earth metals since their hydroxides are strongly basic.
(C) Each element differs from the one above it by the presence of one additional electron in the outer energy level of its atoms.
(D) They are usually stored under water since they react readily with air.
(E) Each has the largest atomic radius of any element in its period.

137.

	E^0
$Li^+(aq) + e^- \rightarrow Li(s)$	-3.02 V
$Na^+(aq) + e^- \rightarrow Na(s)$	-2.71 V

The primary reason for the difference in the standard reduction potentials of Li^+ and Na^+, given above, is that

(A) more energy is required to release an electron from $Na(g)$ than from $Li(g)$
(B) the ionic radius is larger for Li^+ than for Na^+
(C) the electron affinity is larger for Na than for Li
(D) the hydration energy is greater for Na^+ than for Li^+
(E) the hydration energy is greater for Li^+ than for Na^+

138. Which of the following species is diamagnetic? (Atomic numbers: Fe = 26, Co = 27)

(A) ClO_2

(B) NO

(C) N_2O_4

(D) $Fe(CN)_6^{3-}$

(E) $Co(NH_3)_6^{2+}$

139. All of the following are correct statements regarding phosphine, PH_3, EXCEPT:

(A) It is a planar molecule.
(B) It can be prepared by the hydrolysis of Ca_3P_2.
(C) It is a gas at room conditions.
(D) It is a polar molecule.
(E) Its aqueous solutions are less basic than equimolar solutions of NH_3 are.

GO ON TO THE NEXT PAGE.

140. Which of the following is a result of the lanthanide contraction?

(A) Niobium and tantalum are chemically much more similar to each other than either element is to vanadium.
(B) The atomic radius of tungsten is considerably larger than that of molybdenum.
(C) The electropositive character of the copper family elements increases from copper to gold.
(D) The stability of complexes of the tripositive lanthanides decreases from cerium to lutetium.
(E) The lanthanide elements are extremely electropositive metals.

141.

$C_2H_4(g)$: $\Delta H^\circ_{f,298} = 52.3$ kJ/mol
$CO_2(g)$: $\Delta H^\circ_{f,298} = -393.5$ kJ/mol
$H_2O(\ell)$: $\Delta H^\circ_{f,298} = -285.8$ kJ/mol

If the data above are used and if it is assumed that the final products are $CO_2(g)$ and $H_2O(\ell)$, the heat of combustion, ΔH°_{298}, of $C_2H_4(g)$ can be calculated from which of the following?

(A) $-393.5 - 285.8 - 52.3$ kJ/mol
(B) $-393.5 - 285.8 - (-52.3)$ kJ/mol
(C) $2 \times (-393.5) + 2 \times (-285.8) - 52.3$ kJ/mol
(D) $2 \times (-393.5) + 2 \times (-285.8) - (-52.3)$ kJ/mol
(E) $2 \times (393.5) + 2 \times (285.8) - (52.3)$ kJ/mol

GO ON TO THE NEXT PAGE.

142.

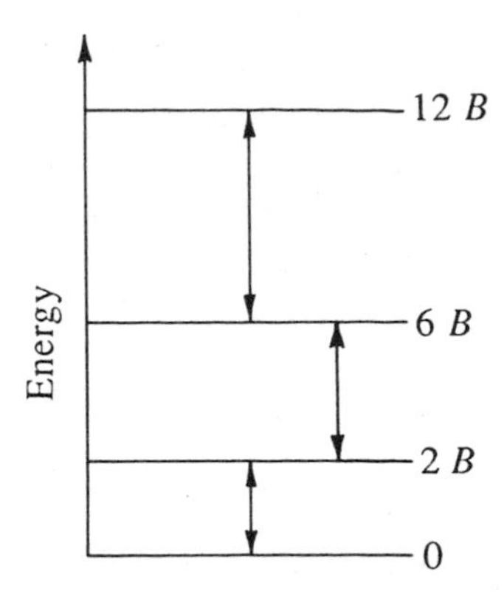

The energy-level diagram shown in the figure above is most appropriately associated with the

(A) vibrational energies of a diatomic molecule
(B) electronic energies of the hydrogen atom
(C) electronic energies of the hydrogen molecule
(D) particle in a one-dimensional box
(E) rotational energies of a diatomic molecule

143. Which of the following must be true for every reaction mixture that exhibits spontaneous reaction? (Q = reaction quotient, K = equilibrium constant, G = Gibbs energy)

(A) $Q < K$
(B) $\Delta G < \Delta G^\circ$
(C) $\Delta G^\circ < \Delta G$
(D) $\Delta G^\circ > 0$
(E) $K < 1$

144. Properties of an ideal gas that are independent of pressure include which of the following?

I. z, collision number
II. λ, mean free path
III. η, coefficient of viscosity

(A) I only
(B) III only
(C) I and II only
(D) I and III only
(E) I, II, and III

145. The average kinetic energy per mole of gaseous molecules at room temperature is most nearly

(A) (3/2) R
(B) (5/2) R
(C) 3 R
(D) (3/2) RT
(E) (5/2) RT

146.

$$CH_3\underset{OH}{CH}-\overset{O}{\overset{\|}{C}}-OH$$

Which of the following bases is a reasonable choice for the resolution of (±)-lactic acid shown above?

(A) $CH_3-\overset{H}{\underset{H}{C}}-NH_2$

(B) $C_6H_5-CH_2-\overset{H}{\underset{H}{C}}-NH_2$

(C) $CH_3C_6H_4-\overset{H}{\underset{H}{C}}-NH_2$

(D) $C_6H_5-\overset{H}{\underset{CH_3}{C}}-NH_2$

(E) $H_2NCH_2-\overset{H}{\underset{CH_3}{C}}-CH_2NH_2$

147. Consideration of the valence bond structures of naphthalene, shown below, should lead one to conclude that which of the indicated bonds is the shortest?

(A) 1
(B) 2
(C) 3
(D) 4
(E) None; all should be of equal length.

GO ON TO THE NEXT PAGE.

148.

1. $(C_6H_5)_3CCl$
2. CH_3Cl
3. $CH_2{=}CH{-}CH_2Cl$
4. $(CH_3)_2C{=}CH{-}CH_2Cl$

Of the following, which lists the compounds above in order of decreasing effectiveness in S_N1 solvolysis reactions?

(A) 1-2-3-4
(B) 1-4-3-2
(C) 4-1-2-3
(D) 4-3-1-2
(E) 4-3-2-1

---

149. From which of the following reactions should one expect very poor (less than 5-10 percent) yields?

(A) $CH_3MgBr + \underset{\diagdown\ O\ \diagup}{CH_2{-}CH_2} \rightarrow CH_3CH_2CH_2OMgBr$

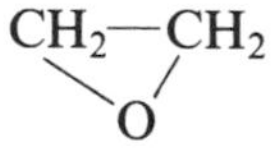

(B) $CH_3MgBr + Br{-}CH_2{-}CH{=}CH_2 \rightarrow CH_3{-}CH_2{-}CH{=}CH_2 + MgBr_2$

(C) $CH_3MgBr + CH_3COOH \rightarrow CH_3C(CH_3)_2OMgBr$

(D) $CH_3MgBr + CH_3COCH_2CH_3 \rightarrow (CH_3)_2C(OMgBr)CH_2CH_3$

(E) $CH_3MgBr + CO_2 \rightarrow CH_3COOMgBr$

GO ON TO THE NEXT PAGE.

150. Which of the following mechanistic steps is LEAST likely to occur?

(A) $CH_3-C(=O)-H + CN^- \longrightarrow CH_3-C(O^-)(CN)-H$

(B) (cyclohexyl)$-\overset{\oplus}{C}H_2 \longrightarrow$ (cyclohexyl cation, $\oplus$)$-CH_3$

(C) cyclohexene $+ Br\cdot \longrightarrow$ cyclohexenyl radical $\cdot + HBr$

(D) (bromo-substituted cyclohexadienyl cation with H and Br, $\oplus$)$-CH_3 + Br^- \longrightarrow$ (cyclohexadiene with Br, CH_3, H, Br)

(E) (benzene)$-CH_2I \xrightarrow[\text{methanol}]{\text{water}}$ (benzene)$-\overset{\oplus}{C}H_2 + I^-$

IF YOU FINISH BEFORE TIME IS CALLED, YOU MAY CHECK YOUR WORK ON THIS TEST.

NO TEST MATERIAL ON THIS PAGE

II

CHEMISTRY TEST

Time—170 minutes

150 Questions

The Subject Tests are intended to measure your achievement in a specialized field of study. Most of the questions are concerned with subject matter that is probably familiar to you, but some of the questions may refer to areas that you have not studied.

Your score will be determined by subtracting one-fourth the number of incorrect answers from the number of correct answers. Questions for which you mark no answer or more than one answer are not counted in scoring. If you have some knowledge of a question and are able to rule out one or more of the answer choices as incorrect, your chances of selecting the correct answer are improved, and answering such questions will likely improve your score. It is unlikely that pure guessing will raise your score; it may lower your score.

You are advised to use your time effectively and to work as rapidly as you can without losing accuracy. Do not spend too much time on questions that are too difficult for you. Go on to the other questions and come back to the difficult ones later if you can.

YOU MUST INDICATE ALL YOUR ANSWERS ON THE SEPARATE ANSWER SHEET. No credit will be given for anything written in this examination book, but to work out your answers you may write in the book as much as you wish. After you have decided which of the suggested answers is best, fill in completely the corresponding space on the answer sheet. Be sure to:

- Use a soft black lead pencil (No. 2 or HB).
- Mark only one answer to each question. No credit will be given for multiple answers.
- Mark your answer in the row with the same number as the number of the question you are answering.
- Carefully and completely fill in the space corresponding to the answer you select for each question. Fill the space with a dark mark so that you cannot see the letter inside the space. Light or partial marks may not be read by the scoring machine. See the example of proper and improper answer marks below.
- Erase all stray marks. If you change an answer, be sure that you completely erase the old answer before marking your new answer. Incomplete erasures may be read as intended answers.

Example:

What city is the capital of France?

(A) Rome
(B) Paris
(C) London
(D) Cairo
(E) Oslo

Sample Answer

Ⓐ ● Ⓒ Ⓓ Ⓔ — CORRECT ANSWER PROPERLY MARKED

Ⓐ Ⓑ Ⓒ Ⓓ Ⓔ
Ⓐ Ⓑ Ⓒ Ⓓ Ⓔ
Ⓐ Ⓑ Ⓒ Ⓓ Ⓔ
Ⓐ Ⓑ Ⓒ Ⓓ Ⓔ — IMPROPER MARKS

Do not be concerned that the answer sheet provides spaces for more answers than there are questions in the test.

CLOSE YOUR TEST BOOK AND WAIT FOR FURTHER INSTRUCTIONS FROM THE SUPERVISOR.

I

NOTE: To ensure the prompt and accurate processing of test results, your cooperation in following these directions is needed. The procedures that follow have been kept to the minimum necessary. They will take a few minutes to complete, but it is essential that you fill in all blanks exactly as directed.

SUBJECT TEST

A. Print and sign your full name in this box:

PRINT: (LAST) (FIRST) (MIDDLE)
SIGN:

B. Side 1 of your answer sheet contains areas that will be used to ensure accurate reporting of your test results. It is essential that you carefully enter the requested information.

1 through 5 YOUR NAME, DATE OF BIRTH, SOCIAL SECURITY NUMBER, REGISTRATION NUMBER, and ADDRESS: Print all the information requested in the boxes and then fill in completely the appropriate oval beneath each entry.

- For date of birth, be sure to enter a zero before a single digit (e.g., if you were born on the third day of the month, you would enter "03" for the day). Use the last two digits of the year of your birth (for 1966, enter 66).
- Copy the registration number from your admission ticket.

6 TITLE CODE: Copy the numbers shown below and fill in completely the appropriate spaces beneath each entry as shown. When you have completed item 6, check to be sure it is identical to the illustration below.

6. TITLE CODE				
2	7	5	3	7
0	0	0	0	0
1	1	1	1	1
●	2	2	2	2
3	3	3	●	3
4	4	4	4	4
5	5	●	5	5
6	6	6	6	6
7	●	7	7	●
8	8	8	8	8
9	9	9	9	9

7 TEST NAME: Copy Chemistry in the box.

FORM CODE: Copy GR9027 in the box.

8 TEST BOOK SERIAL NUMBER: Copy the serial number of your test book in the box. It is printed in red at the upper right on the front cover of your test book.

9 Print the requested information and enter the test center number in the boxes.

10 CERTIFICATION STATEMENT: In the boxed area, WRITE (do not print) the following statement: "I certify that I am the person whose name appears on this answer sheet. I also agree not to disclose the contents of the test I am taking today to anyone." Sign and date where indicated.

C. WHEN YOU HAVE FINISHED THESE INSTRUCTIONS, PLEASE TURN YOUR ANSWER SHEET OVER AND SIGN YOUR NAME IN THE BOX EXACTLY AS YOU DID FOR ITEM 10.

When you have finished, wait for further instructions from the supervisor. DO NOT OPEN YOUR TEST BOOK UNTIL YOU ARE TOLD TO DO SO.

GRADUATE RECORD EXAMINATIONS-SUBJECT TEST

SIDE 1

Use only a pencil with a soft, black lead (No. 2 or HB) to complete this answer sheet.
Be sure to fill in completely the space that corresponds to your answer choice.
Completely erase any errors or stray marks.

DO NOT USE INK.

1. NAME Omit spaces, apostrophes, Jr., II, etc.

Last Name (Family or Surname) - first 15 letters | First Name (Given) - first 12 letters | MI

2. DATE OF BIRTH

Month | Day | Year

Jan. Feb. Mar. Apr. May June July Aug. Sept. Oct. Nov. Dec.

3. SOCIAL SECURITY NUMBER

4. REGISTRATION NUMBER

5. P.O. Box or Street Address first 10 characters

Indicate a space in address by leaving a blank box and filling in the corresponding diamond.

6. TITLE CODE

7. TEST NAME: ______

FORM CODE: ______

8. TEST BOOK SERIAL NUMBER: ______

SHADED AREA FOR ETS USE ONLY

9. YOUR NAME: ______
Last Name(Family or Surname) First Name(Given) M.I.

MAILING ADDRESS: (Print) ______
P.O. Box or Street Address

City State or Province

Country Zip or Postal Code

CENTER: ______
City State or Province

Country Center Number

10. CERTIFICATION STATEMENT

SIGNATURE: ______

DATE: ___/___/___
Month Day Year

540TF30P175e I.N.275416 Q1867-06

SIDE 2

SUBJECT TEST

SIGNATURE:

____

IF YOU DO NOT WANT THIS ANSWER SHEET TO BE SCORED

If you want to cancel your scores from this administration, complete A and B below. No record of this test will be sent to your designated recipients, and there will be no scores on your GRE record. You will receive confirmation of this cancellation. Once a score is canceled, it cannot be reinstated.

To cancel your scores from this test administration, you must:

A. fill in both ovals here . . . . ◯ – ◯

B. sign your name in full below:

____

BE SURE EACH MARK IS DARK AND COMPLETELY FILLS THE INTENDED SPACE AS ILLUSTRATED HERE: ⬬.
YOU MAY FIND MORE RESPONSE SPACES THAN YOU NEED. IF SO, PLEASE LEAVE THEM BLANK.

1. ⒶⒷⒸⒹⒺ	**41.** ⒶⒷⒸⒹⒺ	**81.** ⒶⒷⒸⒹⒺ	**121.** ⒶⒷⒸⒹⒺ	**161.** ⒶⒷⒸⒹⒺ	**201.** ⒶⒷⒸⒹⒺ
2. ⒶⒷⒸⒹⒺ	**42.** ⒶⒷⒸⒹⒺ	**82.** ⒶⒷⒸⒹⒺ	**122.** ⒶⒷⒸⒹⒺ	**162.** ⒶⒷⒸⒹⒺ	**202.** ⒶⒷⒸⒹⒺ
3. ⒶⒷⒸⒹⒺ	**43.** ⒶⒷⒸⒹⒺ	**83.** ⒶⒷⒸⒹⒺ	**123.** ⒶⒷⒸⒹⒺ	**163.** ⒶⒷⒸⒹⒺ	**203.** ⒶⒷⒸⒹⒺ
4. ⒶⒷⒸⒹⒺ	**44.** ⒶⒷⒸⒹⒺ	**84.** ⒶⒷⒸⒹⒺ	**124.** ⒶⒷⒸⒹⒺ	**164.** ⒶⒷⒸⒹⒺ	**204.** ⒶⒷⒸⒹⒺ
5. ⒶⒷⒸⒹⒺ	**45.** ⒶⒷⒸⒹⒺ	**85.** ⒶⒷⒸⒹⒺ	**125.** ⒶⒷⒸⒹⒺ	**165.** ⒶⒷⒸⒹⒺ	**205.** ⒶⒷⒸⒹⒺ
6. ⒶⒷⒸⒹⒺ	**46.** ⒶⒷⒸⒹⒺ	**86.** ⒶⒷⒸⒹⒺ	**126.** ⒶⒷⒸⒹⒺ	**166.** ⒶⒷⒸⒹⒺ	**206.** ⒶⒷⒸⒹⒺ
7. ⒶⒷⒸⒹⒺ	**47.** ⒶⒷⒸⒹⒺ	**87.** ⒶⒷⒸⒹⒺ	**127.** ⒶⒷⒸⒹⒺ	**167.** ⒶⒷⒸⒹⒺ	**207.** ⒶⒷⒸⒹⒺ
8. ⒶⒷⒸⒹⒺ	**48.** ⒶⒷⒸⒹⒺ	**88.** ⒶⒷⒸⒹⒺ	**128.** ⒶⒷⒸⒹⒺ	**168.** ⒶⒷⒸⒹⒺ	**208.** ⒶⒷⒸⒹⒺ
9. ⒶⒷⒸⒹⒺ	**49.** ⒶⒷⒸⒹⒺ	**89.** ⒶⒷⒸⒹⒺ	**129.** ⒶⒷⒸⒹⒺ	**169.** ⒶⒷⒸⒹⒺ	**209.** ⒶⒷⒸⒹⒺ
10. ⒶⒷⒸⒹⒺ	**50.** ⒶⒷⒸⒹⒺ	**90.** ⒶⒷⒸⒹⒺ	**130.** ⒶⒷⒸⒹⒺ	**170.** ⒶⒷⒸⒹⒺ	**210.** ⒶⒷⒸⒹⒺ
11. ⒶⒷⒸⒹⒺ	**51.** ⒶⒷⒸⒹⒺ	**91.** ⒶⒷⒸⒹⒺ	**131.** ⒶⒷⒸⒹⒺ	**171.** ⒶⒷⒸⒹⒺ	**211.** ⒶⒷⒸⒹⒺ
12. ⒶⒷⒸⒹⒺ	**52.** ⒶⒷⒸⒹⒺ	**92.** ⒶⒷⒸⒹⒺ	**132.** ⒶⒷⒸⒹⒺ	**172.** ⒶⒷⒸⒹⒺ	**212.** ⒶⒷⒸⒹⒺ
13. ⒶⒷⒸⒹⒺ	**53.** ⒶⒷⒸⒹⒺ	**93.** ⒶⒷⒸⒹⒺ	**133.** ⒶⒷⒸⒹⒺ	**173.** ⒶⒷⒸⒹⒺ	**213.** ⒶⒷⒸⒹⒺ
14. ⒶⒷⒸⒹⒺ	**54.** ⒶⒷⒸⒹⒺ	**94.** ⒶⒷⒸⒹⒺ	**134.** ⒶⒷⒸⒹⒺ	**174.** ⒶⒷⒸⒹⒺ	**214.** ⒶⒷⒸⒹⒺ
15. ⒶⒷⒸⒹⒺ	**55.** ⒶⒷⒸⒹⒺ	**95.** ⒶⒷⒸⒹⒺ	**135.** ⒶⒷⒸⒹⒺ	**175.** ⒶⒷⒸⒹⒺ	**215.** ⒶⒷⒸⒹⒺ
16. ⒶⒷⒸⒹⒺ	**56.** ⒶⒷⒸⒹⒺ	**96.** ⒶⒷⒸⒹⒺ	**136.** ⒶⒷⒸⒹⒺ	**176.** ⒶⒷⒸⒹⒺ	**216.** ⒶⒷⒸⒹⒺ
17. ⒶⒷⒸⒹⒺ	**57.** ⒶⒷⒸⒹⒺ	**97.** ⒶⒷⒸⒹⒺ	**137.** ⒶⒷⒸⒹⒺ	**177.** ⒶⒷⒸⒹⒺ	**217.** ⒶⒷⒸⒹⒺ
18. ⒶⒷⒸⒹⒺ	**58.** ⒶⒷⒸⒹⒺ	**98.** ⒶⒷⒸⒹⒺ	**138.** ⒶⒷⒸⒹⒺ	**178.** ⒶⒷⒸⒹⒺ	**218.** ⒶⒷⒸⒹⒺ
19. ⒶⒷⒸⒹⒺ	**59.** ⒶⒷⒸⒹⒺ	**99.** ⒶⒷⒸⒹⒺ	**139.** ⒶⒷⒸⒹⒺ	**179.** ⒶⒷⒸⒹⒺ	**219.** ⒶⒷⒸⒹⒺ
20. ⒶⒷⒸⒹⒺ	**60.** ⒶⒷⒸⒹⒺ	**100.** ⒶⒷⒸⒹⒺ	**140.** ⒶⒷⒸⒹⒺ	**180.** ⒶⒷⒸⒹⒺ	**220.** ⒶⒷⒸⒹⒺ
21. ⒶⒷⒸⒹⒺ	**61.** ⒶⒷⒸⒹⒺ	**101.** ⒶⒷⒸⒹⒺ	**141.** ⒶⒷⒸⒹⒺ	**181.** ⒶⒷⒸⒹⒺ	**221.** ⒶⒷⒸⒹⒺ
22. ⒶⒷⒸⒹⒺ	**62.** ⒶⒷⒸⒹⒺ	**102.** ⒶⒷⒸⒹⒺ	**142.** ⒶⒷⒸⒹⒺ	**182.** ⒶⒷⒸⒹⒺ	**222.** ⒶⒷⒸⒹⒺ
23. ⒶⒷⒸⒹⒺ	**63.** ⒶⒷⒸⒹⒺ	**103.** ⒶⒷⒸⒹⒺ	**143.** ⒶⒷⒸⒹⒺ	**183.** ⒶⒷⒸⒹⒺ	**223.** ⒶⒷⒸⒹⒺ
24. ⒶⒷⒸⒹⒺ	**64.** ⒶⒷⒸⒹⒺ	**104.** ⒶⒷⒸⒹⒺ	**144.** ⒶⒷⒸⒹⒺ	**184.** ⒶⒷⒸⒹⒺ	**224.** ⒶⒷⒸⒹⒺ
25. ⒶⒷⒸⒹⒺ	**65.** ⒶⒷⒸⒹⒺ	**105.** ⒶⒷⒸⒹⒺ	**145.** ⒶⒷⒸⒹⒺ	**185.** ⒶⒷⒸⒹⒺ	**225.** ⒶⒷⒸⒹⒺ
26. ⒶⒷⒸⒹⒺ	**66.** ⒶⒷⒸⒹⒺ	**106.** ⒶⒷⒸⒹⒺ	**146.** ⒶⒷⒸⒹⒺ	**186.** ⒶⒷⒸⒹⒺ	**226.** ⒶⒷⒸⒹⒺ
27. ⒶⒷⒸⒹⒺ	**67.** ⒶⒷⒸⒹⒺ	**107.** ⒶⒷⒸⒹⒺ	**147.** ⒶⒷⒸⒹⒺ	**187.** ⒶⒷⒸⒹⒺ	**227.** ⒶⒷⒸⒹⒺ
28. ⒶⒷⒸⒹⒺ	**68.** ⒶⒷⒸⒹⒺ	**108.** ⒶⒷⒸⒹⒺ	**148.** ⒶⒷⒸⒹⒺ	**188.** ⒶⒷⒸⒹⒺ	**228.** ⒶⒷⒸⒹⒺ
29. ⒶⒷⒸⒹⒺ	**69.** ⒶⒷⒸⒹⒺ	**109.** ⒶⒷⒸⒹⒺ	**149.** ⒶⒷⒸⒹⒺ	**189.** ⒶⒷⒸⒹⒺ	**229.** ⒶⒷⒸⒹⒺ
30. ⒶⒷⒸⒹⒺ	**70.** ⒶⒷⒸⒹⒺ	**110.** ⒶⒷⒸⒹⒺ	**150.** ⒶⒷⒸⒹⒺ	**190.** ⒶⒷⒸⒹⒺ	**230.** ⒶⒷⒸⒹⒺ
31. ⒶⒷⒸⒹⒺ	**71.** ⒶⒷⒸⒹⒺ	**111.** ⒶⒷⒸⒹⒺ	**151.** ⒶⒷⒸⒹⒺ	**191.** ⒶⒷⒸⒹⒺ	**231.** ⒶⒷⒸⒹⒺ
32. ⒶⒷⒸⒹⒺ	**72.** ⒶⒷⒸⒹⒺ	**112.** ⒶⒷⒸⒹⒺ	**152.** ⒶⒷⒸⒹⒺ	**192.** ⒶⒷⒸⒹⒺ	**232.** ⒶⒷⒸⒹⒺ
33. ⒶⒷⒸⒹⒺ	**73.** ⒶⒷⒸⒹⒺ	**113.** ⒶⒷⒸⒹⒺ	**153.** ⒶⒷⒸⒹⒺ	**193.** ⒶⒷⒸⒹⒺ	**233.** ⒶⒷⒸⒹⒺ
34. ⒶⒷⒸⒹⒺ	**74.** ⒶⒷⒸⒹⒺ	**114.** ⒶⒷⒸⒹⒺ	**154.** ⒶⒷⒸⒹⒺ	**194.** ⒶⒷⒸⒹⒺ	**234.** ⒶⒷⒸⒹⒺ
35. ⒶⒷⒸⒹⒺ	**75.** ⒶⒷⒸⒹⒺ	**115.** ⒶⒷⒸⒹⒺ	**155.** ⒶⒷⒸⒹⒺ	**195.** ⒶⒷⒸⒹⒺ	**235.** ⒶⒷⒸⒹⒺ
36. ⒶⒷⒸⒹⒺ	**76.** ⒶⒷⒸⒹⒺ	**116.** ⒶⒷⒸⒹⒺ	**156.** ⒶⒷⒸⒹⒺ	**196.** ⒶⒷⒸⒹⒺ	**236.** ⒶⒷⒸⒹⒺ
37. ⒶⒷⒸⒹⒺ	**77.** ⒶⒷⒸⒹⒺ	**117.** ⒶⒷⒸⒹⒺ	**157.** ⒶⒷⒸⒹⒺ	**197.** ⒶⒷⒸⒹⒺ	**237.** ⒶⒷⒸⒹⒺ
38. ⒶⒷⒸⒹⒺ	**78.** ⒶⒷⒸⒹⒺ	**118.** ⒶⒷⒸⒹⒺ	**158.** ⒶⒷⒸⒹⒺ	**198.** ⒶⒷⒸⒹⒺ	**238.** ⒶⒷⒸⒹⒺ
39. ⒶⒷⒸⒹⒺ	**79.** ⒶⒷⒸⒹⒺ	**119.** ⒶⒷⒸⒹⒺ	**159.** ⒶⒷⒸⒹⒺ	**199.** ⒶⒷⒸⒹⒺ	**239.** ⒶⒷⒸⒹⒺ
40. ⒶⒷⒸⒹⒺ	**80.** ⒶⒷⒸⒹⒺ	**120.** ⒶⒷⒸⒹⒺ	**160.** ⒶⒷⒸⒹⒺ	**200.** ⒶⒷⒸⒹⒺ	**240.** ⒶⒷⒸⒹⒺ

FOR ETS USE ONLY	TR	TW	TFS	TCS	1R	1W	1FS	1CS	2R	2W	2FS	2CS	3R	3W	3FS	3CS

GRADUATE RECORD EXAMINATIONS-SUBJECT TEST

SIDE 1

Use only a pencil with a soft, black lead (No. 2 or HB) to complete this answer sheet.
Be sure to fill in completely the space that corresponds to your answer choice.
Completely erase any errors or stray marks.

DO NOT USE INK.

1. NAME Omit spaces, apostrophes, Jr., II, etc.

Last Name (Family or Surname) - first 15 letters | First Name (Given) - first 12 letters | MI

A A A A A A A A A A A A A A A | A A A A A A A A A A A A | A
B B B B B B B B B B B B B B B | B B B B B B B B B B B B | B
C C C C C C C C C C C C C C C | C C C C C C C C C C C C | C
D D D D D D D D D D D D D D D | D D D D D D D D D D D D | D
E E E E E E E E E E E E E E E | E E E E E E E E E E E E | E
F F F F F F F F F F F F F F F | F F F F F F F F F F F F | F
G G G G G G G G G G G G G G G | G G G G G G G G G G G G | G
H H H H H H H H H H H H H H H | H H H H H H H H H H H H | H
I I I I I I I I I I I I I I I | I I I I I I I I I I I I | I
J J J J J J J J J J J J J J J | J J J J J J J J J J J J | J
K K K K K K K K K K K K K K K | K K K K K K K K K K K K | K
L L L L L L L L L L L L L L L | L L L L L L L L L L L L | L
M M M M M M M M M M M M M M M | M M M M M M M M M M M M | M
N N N N N N N N N N N N N N N | N N N N N N N N N N N N | N
O O O O O O O O O O O O O O O | O O O O O O O O O O O O | O
P P P P P P P P P P P P P P P | P P P P P P P P P P P P | P
Q Q Q Q Q Q Q Q Q Q Q Q Q Q Q | Q Q Q Q Q Q Q Q Q Q Q Q | Q
R R R R R R R R R R R R R R R | R R R R R R R R R R R R | R
S S S S S S S S S S S S S S S | S S S S S S S S S S S S | S
T T T T T T T T T T T T T T T | T T T T T T T T T T T T | T
U U U U U U U U U U U U U U U | U U U U U U U U U U U U | U
V V V V V V V V V V V V V V V | V V V V V V V V V V V V | V
W W W W W W W W W W W W W W W | W W W W W W W W W W W W | W
X X X X X X X X X X X X X X X | X X X X X X X X X X X X | X
Y Y Y Y Y Y Y Y Y Y Y Y Y Y Y | Y Y Y Y Y Y Y Y Y Y Y Y | Y
Z Z Z Z Z Z Z Z Z Z Z Z Z Z Z | Z Z Z Z Z Z Z Z Z Z Z Z | Z
– – – – – – – – – – – – – – | – – – – – – – – – – – |

2. DATE OF BIRTH

Month	Day		Year	
Jan.				
Feb.				
Mar.	0	0	0	0
Apr.	1	1	1	1
May	2	2	2	2
June	3	3	3	3
July		4	4	4
Aug.		5	5	5
Sept.		6	6	6
Oct.		7	7	7
Nov.		8		8
Dec.		9		9

3. SOCIAL SECURITY NUMBER

0 0 0 | 0 0 | 0 0 0 0
1 1 1 | 1 1 | 1 1 1 1
2 2 2 | 2 2 | 2 2 2 2
3 3 3 | 3 3 | 3 3 3 3
4 4 4 | 4 4 | 4 4 4 4
5 5 5 | 5 5 | 5 5 5 5
6 6 6 | 6 6 | 6 6 6 6
7 7 7 | 7 7 | 7 7 7 7
8 8 8 | 8 8 | 8 8 8 8
9 9 9 | 9 9 | 9 9 9 9

4. REGISTRATION NUMBER

0 0 0 0 0 0 0
1 1 1 1 1 1 1
2 2 2 2 2 2 2
3 3 3 3 3 3 3
4 4 4 4 4 4 4
5 5 5 5 5 5 5
6 6 6 6 6 6 6
7 7 7 7 7 7 7
8 8 8 8 8 8 8
9 9 9 9 9 9 9

5. P.O. Box or Street Address first 10 characters

Indicate a space in address by leaving a blank box and filling in the corresponding diamond.

◊ ◊ ◊ ◊ ◊ ◊ ◊ ◊ ◊ ◊
A A A A A A A A A A
B B B B B B B B B B
C C C C C C C C C C
D D D D D D D D D D
E E E E E E E E E E
F F F F F F F F F F
G G G G G G G G G G
H H H H H H H H H H
I I I I I I I I I I
J J J J J J J J J J
K K K K K K K K K K
L L L L L L L L L L
M M M M M M M M M M
N N N N N N N N N N
O O O O O O O O O O
P P P P P P P P P P
Q Q Q Q Q Q Q Q Q Q
R R R R R R R R R R
S S S S S S S S S S
T T T T T T T T T T
U U U U U U U U U U
V V V V V V V V V V
W W W W W W W W W W
X X X X X X X X X X
Y Y Y Y Y Y Y Y Y Y
Z Z Z Z Z Z Z Z Z Z
0 0 0 0 0 0 0 0 0 0
1 1 1 1 1 1 1 1 1 1
2 2 2 2 2 2 2 2 2 2
3 3 3 3 3 3 3 3 3 3
4 4 4 4 4 4 4 4 4 4
5 5 5 5 5 5 5 5 5 5
6 6 6 6 6 6 6 6 6 6
7 7 7 7 7 7 7 7 7 7
8 8 8 8 8 8 8 8 8 8
9 9 9 9 9 9 9 9 9 9
/ / / / / / / / / /

6. TITLE CODE

0 0 0 0 0
1 1 1 1 1
2 2 2 2 2
3 3 3 3 3
4 4 4 4 4
5 5 5 5 5
6 6 6 6 6
7 7 7 7 7
8 8 8 8 8
9 9 9 9 9

7. TEST NAME: ______

FORM CODE: ______

8. TEST BOOK SERIAL NUMBER: ______

SHADED AREA FOR ETS USE ONLY

9.

YOUR NAME: ______
Last Name(Family or Surname) First Name(Given) M.I.

MAILING ADDRESS: (Print) ______
P.O. Box or Street Address

______
City State or Province

______
Country Zip or Postal Code

CENTER: ______
City State or Province

______
Country Center Number

10. CERTIFICATION STATEMENT

SIGNATURE: ______

DATE: ____/____/____
Month Day Year

540TF30P175e I.N.275416 Q1867-06

SIDE 2

SUBJECT TEST

SIGNATURE:

____

IF YOU DO NOT WANT THIS ANSWER SHEET TO BE SCORED

If you want to cancel your scores from this administration, complete A and B below. No record of this test will be sent to your designated recipients, and there will be no scores on your GRE record. You will receive confirmation of this cancellation. Once a score is canceled, it cannot be reinstated.

To cancel your scores from this test administration, you must:

A. fill in both ovals here . . . . ◯ – ◯

B. sign your name in full below:

____

BE SURE EACH MARK IS DARK AND COMPLETELY FILLS THE INTENDED SPACE AS ILLUSTRATED HERE: ⬬.
YOU MAY FIND MORE RESPONSE SPACES THAN YOU NEED. IF SO, PLEASE LEAVE THEM BLANK.

1. Ⓐ Ⓑ Ⓒ Ⓓ Ⓔ	41. Ⓐ Ⓑ Ⓒ Ⓓ Ⓔ	81. Ⓐ Ⓑ Ⓒ Ⓓ Ⓔ	121. Ⓐ Ⓑ Ⓒ Ⓓ Ⓔ	161. Ⓐ Ⓑ Ⓒ Ⓓ Ⓔ	201. Ⓐ Ⓑ Ⓒ Ⓓ Ⓔ
2. Ⓐ Ⓑ Ⓒ Ⓓ Ⓔ	42. Ⓐ Ⓑ Ⓒ Ⓓ Ⓔ	82. Ⓐ Ⓑ Ⓒ Ⓓ Ⓔ	122. Ⓐ Ⓑ Ⓒ Ⓓ Ⓔ	162. Ⓐ Ⓑ Ⓒ Ⓓ Ⓔ	202. Ⓐ Ⓑ Ⓒ Ⓓ Ⓔ
3. Ⓐ Ⓑ Ⓒ Ⓓ Ⓔ	43. Ⓐ Ⓑ Ⓒ Ⓓ Ⓔ	83. Ⓐ Ⓑ Ⓒ Ⓓ Ⓔ	123. Ⓐ Ⓑ Ⓒ Ⓓ Ⓔ	163. Ⓐ Ⓑ Ⓒ Ⓓ Ⓔ	203. Ⓐ Ⓑ Ⓒ Ⓓ Ⓔ
4. Ⓐ Ⓑ Ⓒ Ⓓ Ⓔ	44. Ⓐ Ⓑ Ⓒ Ⓓ Ⓔ	84. Ⓐ Ⓑ Ⓒ Ⓓ Ⓔ	124. Ⓐ Ⓑ Ⓒ Ⓓ Ⓔ	164. Ⓐ Ⓑ Ⓒ Ⓓ Ⓔ	204. Ⓐ Ⓑ Ⓒ Ⓓ Ⓔ
5. Ⓐ Ⓑ Ⓒ Ⓓ Ⓔ	45. Ⓐ Ⓑ Ⓒ Ⓓ Ⓔ	85. Ⓐ Ⓑ Ⓒ Ⓓ Ⓔ	125. Ⓐ Ⓑ Ⓒ Ⓓ Ⓔ	165. Ⓐ Ⓑ Ⓒ Ⓓ Ⓔ	205. Ⓐ Ⓑ Ⓒ Ⓓ Ⓔ
6. Ⓐ Ⓑ Ⓒ Ⓓ Ⓔ	46. Ⓐ Ⓑ Ⓒ Ⓓ Ⓔ	86. Ⓐ Ⓑ Ⓒ Ⓓ Ⓔ	126. Ⓐ Ⓑ Ⓒ Ⓓ Ⓔ	166. Ⓐ Ⓑ Ⓒ Ⓓ Ⓔ	206. Ⓐ Ⓑ Ⓒ Ⓓ Ⓔ
7. Ⓐ Ⓑ Ⓒ Ⓓ Ⓔ	47. Ⓐ Ⓑ Ⓒ Ⓓ Ⓔ	87. Ⓐ Ⓑ Ⓒ Ⓓ Ⓔ	127. Ⓐ Ⓑ Ⓒ Ⓓ Ⓔ	167. Ⓐ Ⓑ Ⓒ Ⓓ Ⓔ	207. Ⓐ Ⓑ Ⓒ Ⓓ Ⓔ
8. Ⓐ Ⓑ Ⓒ Ⓓ Ⓔ	48. Ⓐ Ⓑ Ⓒ Ⓓ Ⓔ	88. Ⓐ Ⓑ Ⓒ Ⓓ Ⓔ	128. Ⓐ Ⓑ Ⓒ Ⓓ Ⓔ	168. Ⓐ Ⓑ Ⓒ Ⓓ Ⓔ	208. Ⓐ Ⓑ Ⓒ Ⓓ Ⓔ
9. Ⓐ Ⓑ Ⓒ Ⓓ Ⓔ	49. Ⓐ Ⓑ Ⓒ Ⓓ Ⓔ	89. Ⓐ Ⓑ Ⓒ Ⓓ Ⓔ	129. Ⓐ Ⓑ Ⓒ Ⓓ Ⓔ	169. Ⓐ Ⓑ Ⓒ Ⓓ Ⓔ	209. Ⓐ Ⓑ Ⓒ Ⓓ Ⓔ
10. Ⓐ Ⓑ Ⓒ Ⓓ Ⓔ	50. Ⓐ Ⓑ Ⓒ Ⓓ Ⓔ	90. Ⓐ Ⓑ Ⓒ Ⓓ Ⓔ	130. Ⓐ Ⓑ Ⓒ Ⓓ Ⓔ	170. Ⓐ Ⓑ Ⓒ Ⓓ Ⓔ	210. Ⓐ Ⓑ Ⓒ Ⓓ Ⓔ
11. Ⓐ Ⓑ Ⓒ Ⓓ Ⓔ	51. Ⓐ Ⓑ Ⓒ Ⓓ Ⓔ	91. Ⓐ Ⓑ Ⓒ Ⓓ Ⓔ	131. Ⓐ Ⓑ Ⓒ Ⓓ Ⓔ	171. Ⓐ Ⓑ Ⓒ Ⓓ Ⓔ	211. Ⓐ Ⓑ Ⓒ Ⓓ Ⓔ
12. Ⓐ Ⓑ Ⓒ Ⓓ Ⓔ	52. Ⓐ Ⓑ Ⓒ Ⓓ Ⓔ	92. Ⓐ Ⓑ Ⓒ Ⓓ Ⓔ	132. Ⓐ Ⓑ Ⓒ Ⓓ Ⓔ	172. Ⓐ Ⓑ Ⓒ Ⓓ Ⓔ	212. Ⓐ Ⓑ Ⓒ Ⓓ Ⓔ
13. Ⓐ Ⓑ Ⓒ Ⓓ Ⓔ	53. Ⓐ Ⓑ Ⓒ Ⓓ Ⓔ	93. Ⓐ Ⓑ Ⓒ Ⓓ Ⓔ	133. Ⓐ Ⓑ Ⓒ Ⓓ Ⓔ	173. Ⓐ Ⓑ Ⓒ Ⓓ Ⓔ	213. Ⓐ Ⓑ Ⓒ Ⓓ Ⓔ
14. Ⓐ Ⓑ Ⓒ Ⓓ Ⓔ	54. Ⓐ Ⓑ Ⓒ Ⓓ Ⓔ	94. Ⓐ Ⓑ Ⓒ Ⓓ Ⓔ	134. Ⓐ Ⓑ Ⓒ Ⓓ Ⓔ	174. Ⓐ Ⓑ Ⓒ Ⓓ Ⓔ	214. Ⓐ Ⓑ Ⓒ Ⓓ Ⓔ
15. Ⓐ Ⓑ Ⓒ Ⓓ Ⓔ	55. Ⓐ Ⓑ Ⓒ Ⓓ Ⓔ	95. Ⓐ Ⓑ Ⓒ Ⓓ Ⓔ	135. Ⓐ Ⓑ Ⓒ Ⓓ Ⓔ	175. Ⓐ Ⓑ Ⓒ Ⓓ Ⓔ	215. Ⓐ Ⓑ Ⓒ Ⓓ Ⓔ
16. Ⓐ Ⓑ Ⓒ Ⓓ Ⓔ	56. Ⓐ Ⓑ Ⓒ Ⓓ Ⓔ	96. Ⓐ Ⓑ Ⓒ Ⓓ Ⓔ	136. Ⓐ Ⓑ Ⓒ Ⓓ Ⓔ	176. Ⓐ Ⓑ Ⓒ Ⓓ Ⓔ	216. Ⓐ Ⓑ Ⓒ Ⓓ Ⓔ
17. Ⓐ Ⓑ Ⓒ Ⓓ Ⓔ	57. Ⓐ Ⓑ Ⓒ Ⓓ Ⓔ	97. Ⓐ Ⓑ Ⓒ Ⓓ Ⓔ	137. Ⓐ Ⓑ Ⓒ Ⓓ Ⓔ	177. Ⓐ Ⓑ Ⓒ Ⓓ Ⓔ	217. Ⓐ Ⓑ Ⓒ Ⓓ Ⓔ
18. Ⓐ Ⓑ Ⓒ Ⓓ Ⓔ	58. Ⓐ Ⓑ Ⓒ Ⓓ Ⓔ	98. Ⓐ Ⓑ Ⓒ Ⓓ Ⓔ	138. Ⓐ Ⓑ Ⓒ Ⓓ Ⓔ	178. Ⓐ Ⓑ Ⓒ Ⓓ Ⓔ	218. Ⓐ Ⓑ Ⓒ Ⓓ Ⓔ
19. Ⓐ Ⓑ Ⓒ Ⓓ Ⓔ	59. Ⓐ Ⓑ Ⓒ Ⓓ Ⓔ	99. Ⓐ Ⓑ Ⓒ Ⓓ Ⓔ	139. Ⓐ Ⓑ Ⓒ Ⓓ Ⓔ	179. Ⓐ Ⓑ Ⓒ Ⓓ Ⓔ	219. Ⓐ Ⓑ Ⓒ Ⓓ Ⓔ
20. Ⓐ Ⓑ Ⓒ Ⓓ Ⓔ	60. Ⓐ Ⓑ Ⓒ Ⓓ Ⓔ	100. Ⓐ Ⓑ Ⓒ Ⓓ Ⓔ	140. Ⓐ Ⓑ Ⓒ Ⓓ Ⓔ	180. Ⓐ Ⓑ Ⓒ Ⓓ Ⓔ	220. Ⓐ Ⓑ Ⓒ Ⓓ Ⓔ
21. Ⓐ Ⓑ Ⓒ Ⓓ Ⓔ	61. Ⓐ Ⓑ Ⓒ Ⓓ Ⓔ	101. Ⓐ Ⓑ Ⓒ Ⓓ Ⓔ	141. Ⓐ Ⓑ Ⓒ Ⓓ Ⓔ	181. Ⓐ Ⓑ Ⓒ Ⓓ Ⓔ	221. Ⓐ Ⓑ Ⓒ Ⓓ Ⓔ
22. Ⓐ Ⓑ Ⓒ Ⓓ Ⓔ	62. Ⓐ Ⓑ Ⓒ Ⓓ Ⓔ	102. Ⓐ Ⓑ Ⓒ Ⓓ Ⓔ	142. Ⓐ Ⓑ Ⓒ Ⓓ Ⓔ	182. Ⓐ Ⓑ Ⓒ Ⓓ Ⓔ	222. Ⓐ Ⓑ Ⓒ Ⓓ Ⓔ
23. Ⓐ Ⓑ Ⓒ Ⓓ Ⓔ	63. Ⓐ Ⓑ Ⓒ Ⓓ Ⓔ	103. Ⓐ Ⓑ Ⓒ Ⓓ Ⓔ	143. Ⓐ Ⓑ Ⓒ Ⓓ Ⓔ	183. Ⓐ Ⓑ Ⓒ Ⓓ Ⓔ	223. Ⓐ Ⓑ Ⓒ Ⓓ Ⓔ
24. Ⓐ Ⓑ Ⓒ Ⓓ Ⓔ	64. Ⓐ Ⓑ Ⓒ Ⓓ Ⓔ	104. Ⓐ Ⓑ Ⓒ Ⓓ Ⓔ	144. Ⓐ Ⓑ Ⓒ Ⓓ Ⓔ	184. Ⓐ Ⓑ Ⓒ Ⓓ Ⓔ	224. Ⓐ Ⓑ Ⓒ Ⓓ Ⓔ
25. Ⓐ Ⓑ Ⓒ Ⓓ Ⓔ	65. Ⓐ Ⓑ Ⓒ Ⓓ Ⓔ	105. Ⓐ Ⓑ Ⓒ Ⓓ Ⓔ	145. Ⓐ Ⓑ Ⓒ Ⓓ Ⓔ	185. Ⓐ Ⓑ Ⓒ Ⓓ Ⓔ	225. Ⓐ Ⓑ Ⓒ Ⓓ Ⓔ
26. Ⓐ Ⓑ Ⓒ Ⓓ Ⓔ	66. Ⓐ Ⓑ Ⓒ Ⓓ Ⓔ	106. Ⓐ Ⓑ Ⓒ Ⓓ Ⓔ	146. Ⓐ Ⓑ Ⓒ Ⓓ Ⓔ	186. Ⓐ Ⓑ Ⓒ Ⓓ Ⓔ	226. Ⓐ Ⓑ Ⓒ Ⓓ Ⓔ
27. Ⓐ Ⓑ Ⓒ Ⓓ Ⓔ	67. Ⓐ Ⓑ Ⓒ Ⓓ Ⓔ	107. Ⓐ Ⓑ Ⓒ Ⓓ Ⓔ	147. Ⓐ Ⓑ Ⓒ Ⓓ Ⓔ	187. Ⓐ Ⓑ Ⓒ Ⓓ Ⓔ	227. Ⓐ Ⓑ Ⓒ Ⓓ Ⓔ
28. Ⓐ Ⓑ Ⓒ Ⓓ Ⓔ	68. Ⓐ Ⓑ Ⓒ Ⓓ Ⓔ	108. Ⓐ Ⓑ Ⓒ Ⓓ Ⓔ	148. Ⓐ Ⓑ Ⓒ Ⓓ Ⓔ	188. Ⓐ Ⓑ Ⓒ Ⓓ Ⓔ	228. Ⓐ Ⓑ Ⓒ Ⓓ Ⓔ
29. Ⓐ Ⓑ Ⓒ Ⓓ Ⓔ	69. Ⓐ Ⓑ Ⓒ Ⓓ Ⓔ	109. Ⓐ Ⓑ Ⓒ Ⓓ Ⓔ	149. Ⓐ Ⓑ Ⓒ Ⓓ Ⓔ	189. Ⓐ Ⓑ Ⓒ Ⓓ Ⓔ	229. Ⓐ Ⓑ Ⓒ Ⓓ Ⓔ
30. Ⓐ Ⓑ Ⓒ Ⓓ Ⓔ	70. Ⓐ Ⓑ Ⓒ Ⓓ Ⓔ	110. Ⓐ Ⓑ Ⓒ Ⓓ Ⓔ	150. Ⓐ Ⓑ Ⓒ Ⓓ Ⓔ	190. Ⓐ Ⓑ Ⓒ Ⓓ Ⓔ	230. Ⓐ Ⓑ Ⓒ Ⓓ Ⓔ
31. Ⓐ Ⓑ Ⓒ Ⓓ Ⓔ	71. Ⓐ Ⓑ Ⓒ Ⓓ Ⓔ	111. Ⓐ Ⓑ Ⓒ Ⓓ Ⓔ	151. Ⓐ Ⓑ Ⓒ Ⓓ Ⓔ	191. Ⓐ Ⓑ Ⓒ Ⓓ Ⓔ	231. Ⓐ Ⓑ Ⓒ Ⓓ Ⓔ
32. Ⓐ Ⓑ Ⓒ Ⓓ Ⓔ	72. Ⓐ Ⓑ Ⓒ Ⓓ Ⓔ	112. Ⓐ Ⓑ Ⓒ Ⓓ Ⓔ	152. Ⓐ Ⓑ Ⓒ Ⓓ Ⓔ	192. Ⓐ Ⓑ Ⓒ Ⓓ Ⓔ	232. Ⓐ Ⓑ Ⓒ Ⓓ Ⓔ
33. Ⓐ Ⓑ Ⓒ Ⓓ Ⓔ	73. Ⓐ Ⓑ Ⓒ Ⓓ Ⓔ	113. Ⓐ Ⓑ Ⓒ Ⓓ Ⓔ	153. Ⓐ Ⓑ Ⓒ Ⓓ Ⓔ	193. Ⓐ Ⓑ Ⓒ Ⓓ Ⓔ	233. Ⓐ Ⓑ Ⓒ Ⓓ Ⓔ
34. Ⓐ Ⓑ Ⓒ Ⓓ Ⓔ	74. Ⓐ Ⓑ Ⓒ Ⓓ Ⓔ	114. Ⓐ Ⓑ Ⓒ Ⓓ Ⓔ	154. Ⓐ Ⓑ Ⓒ Ⓓ Ⓔ	194. Ⓐ Ⓑ Ⓒ Ⓓ Ⓔ	234. Ⓐ Ⓑ Ⓒ Ⓓ Ⓔ
35. Ⓐ Ⓑ Ⓒ Ⓓ Ⓔ	75. Ⓐ Ⓑ Ⓒ Ⓓ Ⓔ	115. Ⓐ Ⓑ Ⓒ Ⓓ Ⓔ	155. Ⓐ Ⓑ Ⓒ Ⓓ Ⓔ	195. Ⓐ Ⓑ Ⓒ Ⓓ Ⓔ	235. Ⓐ Ⓑ Ⓒ Ⓓ Ⓔ
36. Ⓐ Ⓑ Ⓒ Ⓓ Ⓔ	76. Ⓐ Ⓑ Ⓒ Ⓓ Ⓔ	116. Ⓐ Ⓑ Ⓒ Ⓓ Ⓔ	156. Ⓐ Ⓑ Ⓒ Ⓓ Ⓔ	196. Ⓐ Ⓑ Ⓒ Ⓓ Ⓔ	236. Ⓐ Ⓑ Ⓒ Ⓓ Ⓔ
37. Ⓐ Ⓑ Ⓒ Ⓓ Ⓔ	77. Ⓐ Ⓑ Ⓒ Ⓓ Ⓔ	117. Ⓐ Ⓑ Ⓒ Ⓓ Ⓔ	157. Ⓐ Ⓑ Ⓒ Ⓓ Ⓔ	197. Ⓐ Ⓑ Ⓒ Ⓓ Ⓔ	237. Ⓐ Ⓑ Ⓒ Ⓓ Ⓔ
38. Ⓐ Ⓑ Ⓒ Ⓓ Ⓔ	78. Ⓐ Ⓑ Ⓒ Ⓓ Ⓔ	118. Ⓐ Ⓑ Ⓒ Ⓓ Ⓔ	158. Ⓐ Ⓑ Ⓒ Ⓓ Ⓔ	198. Ⓐ Ⓑ Ⓒ Ⓓ Ⓔ	238. Ⓐ Ⓑ Ⓒ Ⓓ Ⓔ
39. Ⓐ Ⓑ Ⓒ Ⓓ Ⓔ	79. Ⓐ Ⓑ Ⓒ Ⓓ Ⓔ	119. Ⓐ Ⓑ Ⓒ Ⓓ Ⓔ	159. Ⓐ Ⓑ Ⓒ Ⓓ Ⓔ	199. Ⓐ Ⓑ Ⓒ Ⓓ Ⓔ	239. Ⓐ Ⓑ Ⓒ Ⓓ Ⓔ
40. Ⓐ Ⓑ Ⓒ Ⓓ Ⓔ	80. Ⓐ Ⓑ Ⓒ Ⓓ Ⓔ	120. Ⓐ Ⓑ Ⓒ Ⓓ Ⓔ	160. Ⓐ Ⓑ Ⓒ Ⓓ Ⓔ	200. Ⓐ Ⓑ Ⓒ Ⓓ Ⓔ	240. Ⓐ Ⓑ Ⓒ Ⓓ Ⓔ

FOR ETS USE ONLY	TR	TW	TFS	TCS	1R	1W	1FS	1CS	2R	2W	2FS	2CS	3R	3W	3FS	3CS

GRADUATE RECORD EXAMINATIONS-SUBJECT TEST

SIDE 1

Use only a pencil with a soft, black lead (No. 2 or HB) to complete this answer sheet.
Be sure to fill in completely the space that corresponds to your answer choice.
Completely erase any errors or stray marks.

DO NOT USE INK.

1. NAME Omit spaces, apostrophes, Jr., II, etc.

Last Name (Family or Surname) - first 15 letters | First Name (Given) - first 12 letters | MI

2. DATE OF BIRTH

Month | Day | Year

Jan. Feb. Mar. Apr. May June July Aug. Sept. Oct. Nov. Dec.

3. SOCIAL SECURITY NUMBER

4. REGISTRATION NUMBER

5. P.O. Box or Street Address first 10 characters

Indicate a space in address by leaving a blank box and filling in the corresponding diamond.

6. TITLE CODE

7. TEST NAME: ______

FORM CODE: ______

8. TEST BOOK SERIAL NUMBER: ______

SHADED AREA FOR ETS USE ONLY

9.
YOUR NAME: ______
Last Name(Family or Surname) First Name(Given) M.I.

MAILING ADDRESS: ______
(Print) P.O. Box or Street Address

City State or Province

Country Zip or Postal Code

CENTER: ______
City State or Province

Country Center Number

10. CERTIFICATION STATEMENT

SIGNATURE: ______

DATE: ___/___/___
Month Day Year

540TF30P175e I.N.275416 Q1867-06

SIDE 2

SUBJECT TEST

SIGNATURE:

____

IF YOU DO NOT WANT THIS ANSWER SHEET TO BE SCORED

If you want to cancel your scores from this administration, complete A and B below. No record of this test will be sent to your designated recipients, and there will be no scores on your GRE record. You will receive confirmation of this cancellation. Once a score is canceled, it cannot be reinstated.

To cancel your scores from this test administration, you must:

A. fill in both ovals here . . . . ◯ – ◯

B. sign your name in full below:

____

BE SURE EACH MARK IS DARK AND COMPLETELY FILLS THE INTENDED SPACE AS ILLUSTRATED HERE: ⬬.
YOU MAY FIND MORE RESPONSE SPACES THAN YOU NEED. IF SO, PLEASE LEAVE THEM BLANK.

1. ⒶⒷⒸⒹⒺ	**41.** ⒶⒷⒸⒹⒺ	**81.** ⒶⒷⒸⒹⒺ	**121.** ⒶⒷⒸⒹⒺ	**161.** ⒶⒷⒸⒹⒺ	**201.** ⒶⒷⒸⒹⒺ
2. ⒶⒷⒸⒹⒺ	**42.** ⒶⒷⒸⒹⒺ	**82.** ⒶⒷⒸⒹⒺ	**122.** ⒶⒷⒸⒹⒺ	**162.** ⒶⒷⒸⒹⒺ	**202.** ⒶⒷⒸⒹⒺ
3. ⒶⒷⒸⒹⒺ	**43.** ⒶⒷⒸⒹⒺ	**83.** ⒶⒷⒸⒹⒺ	**123.** ⒶⒷⒸⒹⒺ	**163.** ⒶⒷⒸⒹⒺ	**203.** ⒶⒷⒸⒹⒺ
4. ⒶⒷⒸⒹⒺ	**44.** ⒶⒷⒸⒹⒺ	**84.** ⒶⒷⒸⒹⒺ	**124.** ⒶⒷⒸⒹⒺ	**164.** ⒶⒷⒸⒹⒺ	**204.** ⒶⒷⒸⒹⒺ
5. ⒶⒷⒸⒹⒺ	**45.** ⒶⒷⒸⒹⒺ	**85.** ⒶⒷⒸⒹⒺ	**125.** ⒶⒷⒸⒹⒺ	**165.** ⒶⒷⒸⒹⒺ	**205.** ⒶⒷⒸⒹⒺ
6. ⒶⒷⒸⒹⒺ	**46.** ⒶⒷⒸⒹⒺ	**86.** ⒶⒷⒸⒹⒺ	**126.** ⒶⒷⒸⒹⒺ	**166.** ⒶⒷⒸⒹⒺ	**206.** ⒶⒷⒸⒹⒺ
7. ⒶⒷⒸⒹⒺ	**47.** ⒶⒷⒸⒹⒺ	**87.** ⒶⒷⒸⒹⒺ	**127.** ⒶⒷⒸⒹⒺ	**167.** ⒶⒷⒸⒹⒺ	**207.** ⒶⒷⒸⒹⒺ
8. ⒶⒷⒸⒹⒺ	**48.** ⒶⒷⒸⒹⒺ	**88.** ⒶⒷⒸⒹⒺ	**128.** ⒶⒷⒸⒹⒺ	**168.** ⒶⒷⒸⒹⒺ	**208.** ⒶⒷⒸⒹⒺ
9. ⒶⒷⒸⒹⒺ	**49.** ⒶⒷⒸⒹⒺ	**89.** ⒶⒷⒸⒹⒺ	**129.** ⒶⒷⒸⒹⒺ	**169.** ⒶⒷⒸⒹⒺ	**209.** ⒶⒷⒸⒹⒺ
10. ⒶⒷⒸⒹⒺ	**50.** ⒶⒷⒸⒹⒺ	**90.** ⒶⒷⒸⒹⒺ	**130.** ⒶⒷⒸⒹⒺ	**170.** ⒶⒷⒸⒹⒺ	**210.** ⒶⒷⒸⒹⒺ
11. ⒶⒷⒸⒹⒺ	**51.** ⒶⒷⒸⒹⒺ	**91.** ⒶⒷⒸⒹⒺ	**131.** ⒶⒷⒸⒹⒺ	**171.** ⒶⒷⒸⒹⒺ	**211.** ⒶⒷⒸⒹⒺ
12. ⒶⒷⒸⒹⒺ	**52.** ⒶⒷⒸⒹⒺ	**92.** ⒶⒷⒸⒹⒺ	**132.** ⒶⒷⒸⒹⒺ	**172.** ⒶⒷⒸⒹⒺ	**212.** ⒶⒷⒸⒹⒺ
13. ⒶⒷⒸⒹⒺ	**53.** ⒶⒷⒸⒹⒺ	**93.** ⒶⒷⒸⒹⒺ	**133.** ⒶⒷⒸⒹⒺ	**173.** ⒶⒷⒸⒹⒺ	**213.** ⒶⒷⒸⒹⒺ
14. ⒶⒷⒸⒹⒺ	**54.** ⒶⒷⒸⒹⒺ	**94.** ⒶⒷⒸⒹⒺ	**134.** ⒶⒷⒸⒹⒺ	**174.** ⒶⒷⒸⒹⒺ	**214.** ⒶⒷⒸⒹⒺ
15. ⒶⒷⒸⒹⒺ	**55.** ⒶⒷⒸⒹⒺ	**95.** ⒶⒷⒸⒹⒺ	**135.** ⒶⒷⒸⒹⒺ	**175.** ⒶⒷⒸⒹⒺ	**215.** ⒶⒷⒸⒹⒺ
16. ⒶⒷⒸⒹⒺ	**56.** ⒶⒷⒸⒹⒺ	**96.** ⒶⒷⒸⒹⒺ	**136.** ⒶⒷⒸⒹⒺ	**176.** ⒶⒷⒸⒹⒺ	**216.** ⒶⒷⒸⒹⒺ
17. ⒶⒷⒸⒹⒺ	**57.** ⒶⒷⒸⒹⒺ	**97.** ⒶⒷⒸⒹⒺ	**137.** ⒶⒷⒸⒹⒺ	**177.** ⒶⒷⒸⒹⒺ	**217.** ⒶⒷⒸⒹⒺ
18. ⒶⒷⒸⒹⒺ	**58.** ⒶⒷⒸⒹⒺ	**98.** ⒶⒷⒸⒹⒺ	**138.** ⒶⒷⒸⒹⒺ	**178.** ⒶⒷⒸⒹⒺ	**218.** ⒶⒷⒸⒹⒺ
19. ⒶⒷⒸⒹⒺ	**59.** ⒶⒷⒸⒹⒺ	**99.** ⒶⒷⒸⒹⒺ	**139.** ⒶⒷⒸⒹⒺ	**179.** ⒶⒷⒸⒹⒺ	**219.** ⒶⒷⒸⒹⒺ
20. ⒶⒷⒸⒹⒺ	**60.** ⒶⒷⒸⒹⒺ	**100.** ⒶⒷⒸⒹⒺ	**140.** ⒶⒷⒸⒹⒺ	**180.** ⒶⒷⒸⒹⒺ	**220.** ⒶⒷⒸⒹⒺ
21. ⒶⒷⒸⒹⒺ	**61.** ⒶⒷⒸⒹⒺ	**101.** ⒶⒷⒸⒹⒺ	**141.** ⒶⒷⒸⒹⒺ	**181.** ⒶⒷⒸⒹⒺ	**221.** ⒶⒷⒸⒹⒺ
22. ⒶⒷⒸⒹⒺ	**62.** ⒶⒷⒸⒹⒺ	**102.** ⒶⒷⒸⒹⒺ	**142.** ⒶⒷⒸⒹⒺ	**182.** ⒶⒷⒸⒹⒺ	**222.** ⒶⒷⒸⒹⒺ
23. ⒶⒷⒸⒹⒺ	**63.** ⒶⒷⒸⒹⒺ	**103.** ⒶⒷⒸⒹⒺ	**143.** ⒶⒷⒸⒹⒺ	**183.** ⒶⒷⒸⒹⒺ	**223.** ⒶⒷⒸⒹⒺ
24. ⒶⒷⒸⒹⒺ	**64.** ⒶⒷⒸⒹⒺ	**104.** ⒶⒷⒸⒹⒺ	**144.** ⒶⒷⒸⒹⒺ	**184.** ⒶⒷⒸⒹⒺ	**224.** ⒶⒷⒸⒹⒺ
25. ⒶⒷⒸⒹⒺ	**65.** ⒶⒷⒸⒹⒺ	**105.** ⒶⒷⒸⒹⒺ	**145.** ⒶⒷⒸⒹⒺ	**185.** ⒶⒷⒸⒹⒺ	**225.** ⒶⒷⒸⒹⒺ
26. ⒶⒷⒸⒹⒺ	**66.** ⒶⒷⒸⒹⒺ	**106.** ⒶⒷⒸⒹⒺ	**146.** ⒶⒷⒸⒹⒺ	**186.** ⒶⒷⒸⒹⒺ	**226.** ⒶⒷⒸⒹⒺ
27. ⒶⒷⒸⒹⒺ	**67.** ⒶⒷⒸⒹⒺ	**107.** ⒶⒷⒸⒹⒺ	**147.** ⒶⒷⒸⒹⒺ	**187.** ⒶⒷⒸⒹⒺ	**227.** ⒶⒷⒸⒹⒺ
28. ⒶⒷⒸⒹⒺ	**68.** ⒶⒷⒸⒹⒺ	**108.** ⒶⒷⒸⒹⒺ	**148.** ⒶⒷⒸⒹⒺ	**188.** ⒶⒷⒸⒹⒺ	**228.** ⒶⒷⒸⒹⒺ
29. ⒶⒷⒸⒹⒺ	**69.** ⒶⒷⒸⒹⒺ	**109.** ⒶⒷⒸⒹⒺ	**149.** ⒶⒷⒸⒹⒺ	**189.** ⒶⒷⒸⒹⒺ	**229.** ⒶⒷⒸⒹⒺ
30. ⒶⒷⒸⒹⒺ	**70.** ⒶⒷⒸⒹⒺ	**110.** ⒶⒷⒸⒹⒺ	**150.** ⒶⒷⒸⒹⒺ	**190.** ⒶⒷⒸⒹⒺ	**230.** ⒶⒷⒸⒹⒺ
31. ⒶⒷⒸⒹⒺ	**71.** ⒶⒷⒸⒹⒺ	**111.** ⒶⒷⒸⒹⒺ	**151.** ⒶⒷⒸⒹⒺ	**191.** ⒶⒷⒸⒹⒺ	**231.** ⒶⒷⒸⒹⒺ
32. ⒶⒷⒸⒹⒺ	**72.** ⒶⒷⒸⒹⒺ	**112.** ⒶⒷⒸⒹⒺ	**152.** ⒶⒷⒸⒹⒺ	**192.** ⒶⒷⒸⒹⒺ	**232.** ⒶⒷⒸⒹⒺ
33. ⒶⒷⒸⒹⒺ	**73.** ⒶⒷⒸⒹⒺ	**113.** ⒶⒷⒸⒹⒺ	**153.** ⒶⒷⒸⒹⒺ	**193.** ⒶⒷⒸⒹⒺ	**233.** ⒶⒷⒸⒹⒺ
34. ⒶⒷⒸⒹⒺ	**74.** ⒶⒷⒸⒹⒺ	**114.** ⒶⒷⒸⒹⒺ	**154.** ⒶⒷⒸⒹⒺ	**194.** ⒶⒷⒸⒹⒺ	**234.** ⒶⒷⒸⒹⒺ
35. ⒶⒷⒸⒹⒺ	**75.** ⒶⒷⒸⒹⒺ	**115.** ⒶⒷⒸⒹⒺ	**155.** ⒶⒷⒸⒹⒺ	**195.** ⒶⒷⒸⒹⒺ	**235.** ⒶⒷⒸⒹⒺ
36. ⒶⒷⒸⒹⒺ	**76.** ⒶⒷⒸⒹⒺ	**116.** ⒶⒷⒸⒹⒺ	**156.** ⒶⒷⒸⒹⒺ	**196.** ⒶⒷⒸⒹⒺ	**236.** ⒶⒷⒸⒹⒺ
37. ⒶⒷⒸⒹⒺ	**77.** ⒶⒷⒸⒹⒺ	**117.** ⒶⒷⒸⒹⒺ	**157.** ⒶⒷⒸⒹⒺ	**197.** ⒶⒷⒸⒹⒺ	**237.** ⒶⒷⒸⒹⒺ
38. ⒶⒷⒸⒹⒺ	**78.** ⒶⒷⒸⒹⒺ	**118.** ⒶⒷⒸⒹⒺ	**158.** ⒶⒷⒸⒹⒺ	**198.** ⒶⒷⒸⒹⒺ	**238.** ⒶⒷⒸⒹⒺ
39. ⒶⒷⒸⒹⒺ	**79.** ⒶⒷⒸⒹⒺ	**119.** ⒶⒷⒸⒹⒺ	**159.** ⒶⒷⒸⒹⒺ	**199.** ⒶⒷⒸⒹⒺ	**239.** ⒶⒷⒸⒹⒺ
40. ⒶⒷⒸⒹⒺ	**80.** ⒶⒷⒸⒹⒺ	**120.** ⒶⒷⒸⒹⒺ	**160.** ⒶⒷⒸⒹⒺ	**200.** ⒶⒷⒸⒹⒺ	**240.** ⒶⒷⒸⒹⒺ

FOR ETS USE ONLY	TR	TW	TFS	TCS	1R	1W	1FS	1CS	2R	2W	2FS	2CS	3R	3W	3FS	3CS

GRADUATE RECORD EXAMINATIONS-SUBJECT TEST

SIDE 1

Use only a pencil with a soft, black lead (No. 2 or HB) to complete this answer sheet.
Be sure to fill in completely the space that corresponds to your answer choice.
Completely erase any errors or stray marks.

DO NOT USE INK.

1. NAME Omit spaces, apostrophes, Jr., II, etc.

Last Name (Family or Surname) - first 15 letters | First Name (Given) - first 12 letters | MI

2. DATE OF BIRTH

Month: Jan. Feb. Mar. Apr. May June July Aug. Sept. Oct. Nov. Dec. | Day | Year

3. SOCIAL SECURITY NUMBER

4. REGISTRATION NUMBER

5. P.O. Box or Street Address first 10 characters

Indicate a space in address by leaving a blank box and filling in the corresponding diamond.

6. TITLE CODE

7. TEST NAME: ______

FORM CODE: ______

8. TEST BOOK SERIAL NUMBER: ______

SHADED AREA FOR ETS USE ONLY

9. YOUR NAME: ______
Last Name(Family or Surname) First Name(Given) M.I.

MAILING ADDRESS: (Print) ______
P.O. Box or Street Address

City State or Province

Country Zip or Postal Code

CENTER: ______
City State or Province

Country Center Number

10. CERTIFICATION STATEMENT

SIGNATURE: ______

DATE: ___/___/___
Month Day Year

540TF30P175e I.N.275416 Q1867-06

SIDE 2

SUBJECT TEST

SIGNATURE:

IF YOU DO NOT WANT THIS ANSWER SHEET TO BE SCORED

If you want to cancel your scores from this administration, complete A and B below. No record of this test will be sent to your designated recipients, and there will be no scores on your GRE record. You will receive confirmation of this cancellation. Once a score is canceled, it cannot be reinstated.

To cancel your scores from this test administration, you must:

A. fill in both ovals here . . . . ◯ – ◯

B. sign your name in full below:

BE SURE EACH MARK IS DARK AND COMPLETELY FILLS THE INTENDED SPACE AS ILLUSTRATED HERE: ⬬.
YOU MAY FIND MORE RESPONSE SPACES THAN YOU NEED. IF SO, PLEASE LEAVE THEM BLANK.

1. Ⓐ Ⓑ Ⓒ Ⓓ Ⓔ	**41.** Ⓐ Ⓑ Ⓒ Ⓓ Ⓔ	**81.** Ⓐ Ⓑ Ⓒ Ⓓ Ⓔ	**121.** Ⓐ Ⓑ Ⓒ Ⓓ Ⓔ	**161.** Ⓐ Ⓑ Ⓒ Ⓓ Ⓔ	**201.** Ⓐ Ⓑ Ⓒ Ⓓ Ⓔ
2. Ⓐ Ⓑ Ⓒ Ⓓ Ⓔ	**42.** Ⓐ Ⓑ Ⓒ Ⓓ Ⓔ	**82.** Ⓐ Ⓑ Ⓒ Ⓓ Ⓔ	**122.** Ⓐ Ⓑ Ⓒ Ⓓ Ⓔ	**162.** Ⓐ Ⓑ Ⓒ Ⓓ Ⓔ	**202.** Ⓐ Ⓑ Ⓒ Ⓓ Ⓔ
3. Ⓐ Ⓑ Ⓒ Ⓓ Ⓔ	**43.** Ⓐ Ⓑ Ⓒ Ⓓ Ⓔ	**83.** Ⓐ Ⓑ Ⓒ Ⓓ Ⓔ	**123.** Ⓐ Ⓑ Ⓒ Ⓓ Ⓔ	**163.** Ⓐ Ⓑ Ⓒ Ⓓ Ⓔ	**203.** Ⓐ Ⓑ Ⓒ Ⓓ Ⓔ
4. Ⓐ Ⓑ Ⓒ Ⓓ Ⓔ	**44.** Ⓐ Ⓑ Ⓒ Ⓓ Ⓔ	**84.** Ⓐ Ⓑ Ⓒ Ⓓ Ⓔ	**124.** Ⓐ Ⓑ Ⓒ Ⓓ Ⓔ	**164.** Ⓐ Ⓑ Ⓒ Ⓓ Ⓔ	**204.** Ⓐ Ⓑ Ⓒ Ⓓ Ⓔ
5. Ⓐ Ⓑ Ⓒ Ⓓ Ⓔ	**45.** Ⓐ Ⓑ Ⓒ Ⓓ Ⓔ	**85.** Ⓐ Ⓑ Ⓒ Ⓓ Ⓔ	**125.** Ⓐ Ⓑ Ⓒ Ⓓ Ⓔ	**165.** Ⓐ Ⓑ Ⓒ Ⓓ Ⓔ	**205.** Ⓐ Ⓑ Ⓒ Ⓓ Ⓔ
6. Ⓐ Ⓑ Ⓒ Ⓓ Ⓔ	**46.** Ⓐ Ⓑ Ⓒ Ⓓ Ⓔ	**86.** Ⓐ Ⓑ Ⓒ Ⓓ Ⓔ	**126.** Ⓐ Ⓑ Ⓒ Ⓓ Ⓔ	**166.** Ⓐ Ⓑ Ⓒ Ⓓ Ⓔ	**206.** Ⓐ Ⓑ Ⓒ Ⓓ Ⓔ
7. Ⓐ Ⓑ Ⓒ Ⓓ Ⓔ	**47.** Ⓐ Ⓑ Ⓒ Ⓓ Ⓔ	**87.** Ⓐ Ⓑ Ⓒ Ⓓ Ⓔ	**127.** Ⓐ Ⓑ Ⓒ Ⓓ Ⓔ	**167.** Ⓐ Ⓑ Ⓒ Ⓓ Ⓔ	**207.** Ⓐ Ⓑ Ⓒ Ⓓ Ⓔ
8. Ⓐ Ⓑ Ⓒ Ⓓ Ⓔ	**48.** Ⓐ Ⓑ Ⓒ Ⓓ Ⓔ	**88.** Ⓐ Ⓑ Ⓒ Ⓓ Ⓔ	**128.** Ⓐ Ⓑ Ⓒ Ⓓ Ⓔ	**168.** Ⓐ Ⓑ Ⓒ Ⓓ Ⓔ	**208.** Ⓐ Ⓑ Ⓒ Ⓓ Ⓔ
9. Ⓐ Ⓑ Ⓒ Ⓓ Ⓔ	**49.** Ⓐ Ⓑ Ⓒ Ⓓ Ⓔ	**89.** Ⓐ Ⓑ Ⓒ Ⓓ Ⓔ	**129.** Ⓐ Ⓑ Ⓒ Ⓓ Ⓔ	**169.** Ⓐ Ⓑ Ⓒ Ⓓ Ⓔ	**209.** Ⓐ Ⓑ Ⓒ Ⓓ Ⓔ
10. Ⓐ Ⓑ Ⓒ Ⓓ Ⓔ	**50.** Ⓐ Ⓑ Ⓒ Ⓓ Ⓔ	**90.** Ⓐ Ⓑ Ⓒ Ⓓ Ⓔ	**130.** Ⓐ Ⓑ Ⓒ Ⓓ Ⓔ	**170.** Ⓐ Ⓑ Ⓒ Ⓓ Ⓔ	**210.** Ⓐ Ⓑ Ⓒ Ⓓ Ⓔ
11. Ⓐ Ⓑ Ⓒ Ⓓ Ⓔ	**51.** Ⓐ Ⓑ Ⓒ Ⓓ Ⓔ	**91.** Ⓐ Ⓑ Ⓒ Ⓓ Ⓔ	**131.** Ⓐ Ⓑ Ⓒ Ⓓ Ⓔ	**171.** Ⓐ Ⓑ Ⓒ Ⓓ Ⓔ	**211.** Ⓐ Ⓑ Ⓒ Ⓓ Ⓔ
12. Ⓐ Ⓑ Ⓒ Ⓓ Ⓔ	**52.** Ⓐ Ⓑ Ⓒ Ⓓ Ⓔ	**92.** Ⓐ Ⓑ Ⓒ Ⓓ Ⓔ	**132.** Ⓐ Ⓑ Ⓒ Ⓓ Ⓔ	**172.** Ⓐ Ⓑ Ⓒ Ⓓ Ⓔ	**212.** Ⓐ Ⓑ Ⓒ Ⓓ Ⓔ
13. Ⓐ Ⓑ Ⓒ Ⓓ Ⓔ	**53.** Ⓐ Ⓑ Ⓒ Ⓓ Ⓔ	**93.** Ⓐ Ⓑ Ⓒ Ⓓ Ⓔ	**133.** Ⓐ Ⓑ Ⓒ Ⓓ Ⓔ	**173.** Ⓐ Ⓑ Ⓒ Ⓓ Ⓔ	**213.** Ⓐ Ⓑ Ⓒ Ⓓ Ⓔ
14. Ⓐ Ⓑ Ⓒ Ⓓ Ⓔ	**54.** Ⓐ Ⓑ Ⓒ Ⓓ Ⓔ	**94.** Ⓐ Ⓑ Ⓒ Ⓓ Ⓔ	**134.** Ⓐ Ⓑ Ⓒ Ⓓ Ⓔ	**174.** Ⓐ Ⓑ Ⓒ Ⓓ Ⓔ	**214.** Ⓐ Ⓑ Ⓒ Ⓓ Ⓔ
15. Ⓐ Ⓑ Ⓒ Ⓓ Ⓔ	**55.** Ⓐ Ⓑ Ⓒ Ⓓ Ⓔ	**95.** Ⓐ Ⓑ Ⓒ Ⓓ Ⓔ	**135.** Ⓐ Ⓑ Ⓒ Ⓓ Ⓔ	**175.** Ⓐ Ⓑ Ⓒ Ⓓ Ⓔ	**215.** Ⓐ Ⓑ Ⓒ Ⓓ Ⓔ
16. Ⓐ Ⓑ Ⓒ Ⓓ Ⓔ	**56.** Ⓐ Ⓑ Ⓒ Ⓓ Ⓔ	**96.** Ⓐ Ⓑ Ⓒ Ⓓ Ⓔ	**136.** Ⓐ Ⓑ Ⓒ Ⓓ Ⓔ	**176.** Ⓐ Ⓑ Ⓒ Ⓓ Ⓔ	**216.** Ⓐ Ⓑ Ⓒ Ⓓ Ⓔ
17. Ⓐ Ⓑ Ⓒ Ⓓ Ⓔ	**57.** Ⓐ Ⓑ Ⓒ Ⓓ Ⓔ	**97.** Ⓐ Ⓑ Ⓒ Ⓓ Ⓔ	**137.** Ⓐ Ⓑ Ⓒ Ⓓ Ⓔ	**177.** Ⓐ Ⓑ Ⓒ Ⓓ Ⓔ	**217.** Ⓐ Ⓑ Ⓒ Ⓓ Ⓔ
18. Ⓐ Ⓑ Ⓒ Ⓓ Ⓔ	**58.** Ⓐ Ⓑ Ⓒ Ⓓ Ⓔ	**98.** Ⓐ Ⓑ Ⓒ Ⓓ Ⓔ	**138.** Ⓐ Ⓑ Ⓒ Ⓓ Ⓔ	**178.** Ⓐ Ⓑ Ⓒ Ⓓ Ⓔ	**218.** Ⓐ Ⓑ Ⓒ Ⓓ Ⓔ
19. Ⓐ Ⓑ Ⓒ Ⓓ Ⓔ	**59.** Ⓐ Ⓑ Ⓒ Ⓓ Ⓔ	**99.** Ⓐ Ⓑ Ⓒ Ⓓ Ⓔ	**139.** Ⓐ Ⓑ Ⓒ Ⓓ Ⓔ	**179.** Ⓐ Ⓑ Ⓒ Ⓓ Ⓔ	**219.** Ⓐ Ⓑ Ⓒ Ⓓ Ⓔ
20. Ⓐ Ⓑ Ⓒ Ⓓ Ⓔ	**60.** Ⓐ Ⓑ Ⓒ Ⓓ Ⓔ	**100.** Ⓐ Ⓑ Ⓒ Ⓓ Ⓔ	**140.** Ⓐ Ⓑ Ⓒ Ⓓ Ⓔ	**180.** Ⓐ Ⓑ Ⓒ Ⓓ Ⓔ	**220.** Ⓐ Ⓑ Ⓒ Ⓓ Ⓔ
21. Ⓐ Ⓑ Ⓒ Ⓓ Ⓔ	**61.** Ⓐ Ⓑ Ⓒ Ⓓ Ⓔ	**101.** Ⓐ Ⓑ Ⓒ Ⓓ Ⓔ	**141.** Ⓐ Ⓑ Ⓒ Ⓓ Ⓔ	**181.** Ⓐ Ⓑ Ⓒ Ⓓ Ⓔ	**221.** Ⓐ Ⓑ Ⓒ Ⓓ Ⓔ
22. Ⓐ Ⓑ Ⓒ Ⓓ Ⓔ	**62.** Ⓐ Ⓑ Ⓒ Ⓓ Ⓔ	**102.** Ⓐ Ⓑ Ⓒ Ⓓ Ⓔ	**142.** Ⓐ Ⓑ Ⓒ Ⓓ Ⓔ	**182.** Ⓐ Ⓑ Ⓒ Ⓓ Ⓔ	**222.** Ⓐ Ⓑ Ⓒ Ⓓ Ⓔ
23. Ⓐ Ⓑ Ⓒ Ⓓ Ⓔ	**63.** Ⓐ Ⓑ Ⓒ Ⓓ Ⓔ	**103.** Ⓐ Ⓑ Ⓒ Ⓓ Ⓔ	**143.** Ⓐ Ⓑ Ⓒ Ⓓ Ⓔ	**183.** Ⓐ Ⓑ Ⓒ Ⓓ Ⓔ	**223.** Ⓐ Ⓑ Ⓒ Ⓓ Ⓔ
24. Ⓐ Ⓑ Ⓒ Ⓓ Ⓔ	**64.** Ⓐ Ⓑ Ⓒ Ⓓ Ⓔ	**104.** Ⓐ Ⓑ Ⓒ Ⓓ Ⓔ	**144.** Ⓐ Ⓑ Ⓒ Ⓓ Ⓔ	**184.** Ⓐ Ⓑ Ⓒ Ⓓ Ⓔ	**224.** Ⓐ Ⓑ Ⓒ Ⓓ Ⓔ
25. Ⓐ Ⓑ Ⓒ Ⓓ Ⓔ	**65.** Ⓐ Ⓑ Ⓒ Ⓓ Ⓔ	**105.** Ⓐ Ⓑ Ⓒ Ⓓ Ⓔ	**145.** Ⓐ Ⓑ Ⓒ Ⓓ Ⓔ	**185.** Ⓐ Ⓑ Ⓒ Ⓓ Ⓔ	**225.** Ⓐ Ⓑ Ⓒ Ⓓ Ⓔ
26. Ⓐ Ⓑ Ⓒ Ⓓ Ⓔ	**66.** Ⓐ Ⓑ Ⓒ Ⓓ Ⓔ	**106.** Ⓐ Ⓑ Ⓒ Ⓓ Ⓔ	**146.** Ⓐ Ⓑ Ⓒ Ⓓ Ⓔ	**186.** Ⓐ Ⓑ Ⓒ Ⓓ Ⓔ	**226.** Ⓐ Ⓑ Ⓒ Ⓓ Ⓔ
27. Ⓐ Ⓑ Ⓒ Ⓓ Ⓔ	**67.** Ⓐ Ⓑ Ⓒ Ⓓ Ⓔ	**107.** Ⓐ Ⓑ Ⓒ Ⓓ Ⓔ	**147.** Ⓐ Ⓑ Ⓒ Ⓓ Ⓔ	**187.** Ⓐ Ⓑ Ⓒ Ⓓ Ⓔ	**227.** Ⓐ Ⓑ Ⓒ Ⓓ Ⓔ
28. Ⓐ Ⓑ Ⓒ Ⓓ Ⓔ	**68.** Ⓐ Ⓑ Ⓒ Ⓓ Ⓔ	**108.** Ⓐ Ⓑ Ⓒ Ⓓ Ⓔ	**148.** Ⓐ Ⓑ Ⓒ Ⓓ Ⓔ	**188.** Ⓐ Ⓑ Ⓒ Ⓓ Ⓔ	**228.** Ⓐ Ⓑ Ⓒ Ⓓ Ⓔ
29. Ⓐ Ⓑ Ⓒ Ⓓ Ⓔ	**69.** Ⓐ Ⓑ Ⓒ Ⓓ Ⓔ	**109.** Ⓐ Ⓑ Ⓒ Ⓓ Ⓔ	**149.** Ⓐ Ⓑ Ⓒ Ⓓ Ⓔ	**189.** Ⓐ Ⓑ Ⓒ Ⓓ Ⓔ	**229.** Ⓐ Ⓑ Ⓒ Ⓓ Ⓔ
30. Ⓐ Ⓑ Ⓒ Ⓓ Ⓔ	**70.** Ⓐ Ⓑ Ⓒ Ⓓ Ⓔ	**110.** Ⓐ Ⓑ Ⓒ Ⓓ Ⓔ	**150.** Ⓐ Ⓑ Ⓒ Ⓓ Ⓔ	**190.** Ⓐ Ⓑ Ⓒ Ⓓ Ⓔ	**230.** Ⓐ Ⓑ Ⓒ Ⓓ Ⓔ
31. Ⓐ Ⓑ Ⓒ Ⓓ Ⓔ	**71.** Ⓐ Ⓑ Ⓒ Ⓓ Ⓔ	**111.** Ⓐ Ⓑ Ⓒ Ⓓ Ⓔ	**151.** Ⓐ Ⓑ Ⓒ Ⓓ Ⓔ	**191.** Ⓐ Ⓑ Ⓒ Ⓓ Ⓔ	**231.** Ⓐ Ⓑ Ⓒ Ⓓ Ⓔ
32. Ⓐ Ⓑ Ⓒ Ⓓ Ⓔ	**72.** Ⓐ Ⓑ Ⓒ Ⓓ Ⓔ	**112.** Ⓐ Ⓑ Ⓒ Ⓓ Ⓔ	**152.** Ⓐ Ⓑ Ⓒ Ⓓ Ⓔ	**192.** Ⓐ Ⓑ Ⓒ Ⓓ Ⓔ	**232.** Ⓐ Ⓑ Ⓒ Ⓓ Ⓔ
33. Ⓐ Ⓑ Ⓒ Ⓓ Ⓔ	**73.** Ⓐ Ⓑ Ⓒ Ⓓ Ⓔ	**113.** Ⓐ Ⓑ Ⓒ Ⓓ Ⓔ	**153.** Ⓐ Ⓑ Ⓒ Ⓓ Ⓔ	**193.** Ⓐ Ⓑ Ⓒ Ⓓ Ⓔ	**233.** Ⓐ Ⓑ Ⓒ Ⓓ Ⓔ
34. Ⓐ Ⓑ Ⓒ Ⓓ Ⓔ	**74.** Ⓐ Ⓑ Ⓒ Ⓓ Ⓔ	**114.** Ⓐ Ⓑ Ⓒ Ⓓ Ⓔ	**154.** Ⓐ Ⓑ Ⓒ Ⓓ Ⓔ	**194.** Ⓐ Ⓑ Ⓒ Ⓓ Ⓔ	**234.** Ⓐ Ⓑ Ⓒ Ⓓ Ⓔ
35. Ⓐ Ⓑ Ⓒ Ⓓ Ⓔ	**75.** Ⓐ Ⓑ Ⓒ Ⓓ Ⓔ	**115.** Ⓐ Ⓑ Ⓒ Ⓓ Ⓔ	**155.** Ⓐ Ⓑ Ⓒ Ⓓ Ⓔ	**195.** Ⓐ Ⓑ Ⓒ Ⓓ Ⓔ	**235.** Ⓐ Ⓑ Ⓒ Ⓓ Ⓔ
36. Ⓐ Ⓑ Ⓒ Ⓓ Ⓔ	**76.** Ⓐ Ⓑ Ⓒ Ⓓ Ⓔ	**116.** Ⓐ Ⓑ Ⓒ Ⓓ Ⓔ	**156.** Ⓐ Ⓑ Ⓒ Ⓓ Ⓔ	**196.** Ⓐ Ⓑ Ⓒ Ⓓ Ⓔ	**236.** Ⓐ Ⓑ Ⓒ Ⓓ Ⓔ
37. Ⓐ Ⓑ Ⓒ Ⓓ Ⓔ	**77.** Ⓐ Ⓑ Ⓒ Ⓓ Ⓔ	**117.** Ⓐ Ⓑ Ⓒ Ⓓ Ⓔ	**157.** Ⓐ Ⓑ Ⓒ Ⓓ Ⓔ	**197.** Ⓐ Ⓑ Ⓒ Ⓓ Ⓔ	**237.** Ⓐ Ⓑ Ⓒ Ⓓ Ⓔ
38. Ⓐ Ⓑ Ⓒ Ⓓ Ⓔ	**78.** Ⓐ Ⓑ Ⓒ Ⓓ Ⓔ	**118.** Ⓐ Ⓑ Ⓒ Ⓓ Ⓔ	**158.** Ⓐ Ⓑ Ⓒ Ⓓ Ⓔ	**198.** Ⓐ Ⓑ Ⓒ Ⓓ Ⓔ	**238.** Ⓐ Ⓑ Ⓒ Ⓓ Ⓔ
39. Ⓐ Ⓑ Ⓒ Ⓓ Ⓔ	**79.** Ⓐ Ⓑ Ⓒ Ⓓ Ⓔ	**119.** Ⓐ Ⓑ Ⓒ Ⓓ Ⓔ	**159.** Ⓐ Ⓑ Ⓒ Ⓓ Ⓔ	**199.** Ⓐ Ⓑ Ⓒ Ⓓ Ⓔ	**239.** Ⓐ Ⓑ Ⓒ Ⓓ Ⓔ
40. Ⓐ Ⓑ Ⓒ Ⓓ Ⓔ	**80.** Ⓐ Ⓑ Ⓒ Ⓓ Ⓔ	**120.** Ⓐ Ⓑ Ⓒ Ⓓ Ⓔ	**160.** Ⓐ Ⓑ Ⓒ Ⓓ Ⓔ	**200.** Ⓐ Ⓑ Ⓒ Ⓓ Ⓔ	**240.** Ⓐ Ⓑ Ⓒ Ⓓ Ⓔ

FOR ETS USE ONLY	TR	TW	TFS	TCS	1R	1W	1FS	1CS	2R	2W	2FS	2CS	3R	3W	3FS	3CS

GRE® PUBLICATIONS ORDER FORM
1990-91

Graduate Record Examinations
Educational Testing Service
P.O. Box 6014
Princeton, NJ 08541-6014

Item Number	Publication	Price*	No. of Copies	Amount	Total
	Practice Test Books (540-01)				
241245	†Practicing to Take the GRE General Test—No. 8	$12.00			
241235	Practicing to Take the GRE General Test—No. 7	10.00			
241241	†Practicing to Take the GRE Biology Test–2nd Edition	11.00			
241242	†Practicing to Take the GRE Chemistry Test–2nd Edition	11.00			
241247	Practicing to Take the GRE Computer Science Test	9.00			
241218	Practicing to Take the GRE Economics Test	9.00			
241236	Practicing to Take the GRE Education Test–2nd Edition	11.00			
241237	Practicing to Take the GRE Engineering Test–2nd Edition	11.00			
241227	Practicing to Take the GRE Geology Test	9.00			
241219	Practicing to Take the GRE History Test	9.00			
241243	†Practicing to Take the GRE Literature in English Test—2nd Edition	11.00			
241228	Practicing to Take the GRE Mathematics Test	9.00			
241244	Practicing to Take the GRE Revised Music Test	9.00			
241246	Practicing to Take the GRE Physics Test	9.00			
241234	Practicing to Take the GRE Political Science Test	9.00			
241238	Practicing to Take the GRE Psychology Test–2nd Edition	11.00			
241229	Practicing to Take the GRE Sociology Test	9.00			
	Software Editions (540-07)				
299623	†Practicing to Take the GRE General Test—No. 7, Apple Macintosh Software Edition Version 2.1	70.00			
299624	†Practicing to Take the GRE General Test—No. 7, IBM Software Edition Version 2.1	70.00			
	Directory of Graduate Programs (540-99)				
252025	Volume A—Agriculture, Biological Sciences, Psychology, Health Sciences, and Home Economics	14.00			
252026	Volume B—Arts and Humanities	14.00			
252027	Volume C—Physical Sciences, Mathematics, and Engineering	14.00			
252028	Volume D—Social Sciences and Education	14.00			

* **Postage: In North America, U.S. Territories, and APO addresses,** postage and handling to a single address is included in the price of the publication.
To all other locations (airmail only) for postage and handling to a single address, add $4 for the first book ordered and $2 for each additional book. Add $10 for each software edition ordered.

† Available September 1990.

POSTAGE 540-52

TOTAL AMOUNT ENCLOSED

- Allow three to four weeks for delivery.
- Payment should be made by check or money order drawn on a U.S. bank, U.S. Postal Money Order, UNESCO Coupons, or current International Postal Reply Coupons.
- *Orders received without payment or a purchase order will be returned.*

Make your remittance payable to Graduate Record Examinations.

ETS use only

TYPE OR PRINT CLEARLY BELOW. DO NOT DETACH THESE MAILING LABELS.

Graduate Record Examinations
Educational Testing Service
P.O. Box 6014
Princeton, NJ 08541-6014

TO: ______________________

____

____

____

Graduate Record Examinations
Educational Testing Service
P.O. Box 6014
Princeton, NJ 08541-6014

TO: ______________________

____

____

____